Climate Change Temporalities

Climate Change Temporalities explores how various timescales, timespans, intervals, rhythms, cycles, and changes in acceleration are at play in climate change discourses. It argues that nuanced, detailed, and specific understandings and concepts are required to handle the challenges of a climatically changed world, politically and socially as well as scientifically. Rather than reflecting abstractly on theories of temporality, this edited collection explores a variety of timescales and temporalities from narratives, experience, popular culture, and everyday life in addition to science and history – and the entanglements between them. The chapters are clustered into three main sections, exploring a range of genres, such as questionnaires, interviews, magazines, news media, television series, aquariums, and popular science books to critically examine how and where climate change understandings are formed. The book also includes chapters historising notions of climate and temporality by exploring scientific debates and practices.

Climate Change Temporalities will be of great interest to students and scholars of humanistic climate change research, environmental humanities, studies of temporality and historicity, cultural studies, cultural history, and popular culture.

Kyrre Kverndokk is Professor of Cultural Studies at the University of Bergen, Norway.

Marit Ruge Bjærke is a Postdoctoral Research Fellow in Cultural Studies at the University of Bergen, Norway.

Anne Eriksen is Professor of Cultural History at the University of Oslo, Norway.

Routledge Explorations in Environmental Studies

The Environmental Impact of Overpopulation
The Ethics of Procreation
Trevor Hedberg

Riverlands of the Anthropocene
Walking Our Waterways as Places of Becoming
Margaret Somerville

An Environmental History of Australian Rainforests until 1939
Fire, Rain, Settlers and Conservation
Warwick Frost

Daoism and Environmental Philosophy
Nourishing Life
Eric S. Nelson

Ecological Law and the Planetary Crisis
A Legal Guide for Harmony on Earth
Geoffrey Garver

From Environmental to Ecological Law
Kirsten Anker; Peter D Burdon; Geoffrey Garver; Michelle Maloney and Carla Sbert

Climate Change Temporalities
Explorations in Vernacular, Popular, and Scientific Discourse
Edited by Kyrre Kverndokk, Marit Ruge Bjærke, and Anne Eriksen

https://www.routledge.com/Routledge-Explorations-in-Environmental-Studies/book-series/REES

Climate Change Temporalities

Explorations in Vernacular, Popular, and Scientific Discourse

Edited by
Kyrre Kverndokk,
Marit Ruge Bjærke, and
Anne Eriksen

LONDON AND NEW YORK

First published 2021
by Routledge
2 Park Square, Milton Park, Abingdon, Oxon OX14 4RN

and by Routledge
52 Vanderbilt Avenue, New York, NY 10017

Routledge is an imprint of the Taylor & Francis Group, an informa business

British Library Cataloguing-in-Publication Data
A catalogue record for this book is available from the British Library

Library of Congress Cataloging-in-Publication Data
A catalog record has been requested for this book

ISBN: 978-0-367-47960-2 (hbk)
ISBN: 978-0-367-69640-5 (pbk)
ISBN: 978-1-003-03741-5 (ebk)

Typeset in Bembo
by codeMantra

Contents

Contributors

Marit Ruge Bjærke is a postdoctoral research fellow in Cultural Studies at the Department of Archaeology, History, Cultural Studies, and Religion, University of Bergen, Norway. Bjærke holds a PhD in Marine Biology and an MA in History of Ideas. Her research interests lie within the environmental humanities, with a focus on biodiversity, biodiversity loss, and invasive species.

Anne Eriksen is Professor of Cultural History at the Department of Culture Studies and Oriental Languages, University of Oslo, Norway. Her research interests include collective memory and folklore, cultural heritage, the history of knowledge, and eighteenth-century historiography and medical history.

Diane E. Goldstein is Professor of Folklore Studies at the Department of Folklore and Ethnomusicology at Indiana University, USA. Her research interests are vernacular narrative, belief studies, and ethnography of speaking as well as the folklore of violence, trauma, and illness.

Isak Winkel Holm is Professor of Comparative Literature at the Department of Arts and Cultural Studies, University of Copenhagen. His interests are literary theory, philosophical aesthetics, political theory, and disaster research. He is currently working on a monograph on Søren Kierkegaard and the ongoing ecological catastrophe.

Camilla Asplund Ingemark is a research fellow (Researcher II) at the Department of Archaeology, History, Cultural Studies, and Religion at the University of Bergen, Norway, for 2017–2020. She is also Senior Lecturer in Ethnology at Uppsala University, Campus Gotland, Sweden, and Docent in Folkloristics at Åbo Akademi University, Finland. Her research interests include climate change temporalities, vernacular conceptions of history, the cultural history of disasters, folk narrative from antiquity to the present, and the history of emotions.

Lars Kaijser is Professor of Ethnology at the Department for Ethnology, History of Religion, and Gender Studies at Stockholm University, Sweden.

His research interests are popular music heritage, guided tours, and the intersection of nature/culture.

Kyrre Kverndokk is Professor of Cultural Studies at the Department of Archaeology, History, Cultural Studies, and Religion at the University of Bergen, Norway. His research interests are the history of folklore studies, the cultural history of natural disasters, climate change temporalities, and the practice and politics of Second World War memory.

Lone Ree Milkær is a PhD student at the Department of Archaeology, History, Cultural Studies, and Religion at the University of Bergen, Norway. Her research interests are climate change discourse, everyday climate activism, traditionalisation, and environmental humanities.

John Ødemark is Professor of Cultural History and Cultural Encounters at the Department of Culture Studies and Oriental Languages at the University of Oslo, Norway. His research interests are the history of the humanities and cultural translation.

Illustrations

Acknowledgements

This book is one of the outcomes of the research project 'The Future is Now: Temporality and Exemplarity in Climate Change Discourses', which started in 2017. It was funded by the Research Council of Norway's Large-scale Programme on Climate Research and has been located in the Department of Archaeology, History, Cultural Studies, and Religion at the University of Bergen, Norway. We would like to express our thanks to both our department and the Faculty of Humanities for their support and commitment to promoting humanistic climate change research. Discussing our work in the faculty research group in Environmental Humanities has been both fruitful and encouraging.

The project has involved scholars from the University of Oslo, Uppsala University, Copenhagen University, Stockholm University, and Indiana University. As one of the aims of this project has been to confront established methodological and theoretical approaches from the humanities with new sets of problems and approaches from environmental studies, a large degree of interdisciplinarity has been necessary. In addition to the contributors in the book, geologist Henrik H. Svensen has been one of the core members of the project team. With a special competence in paleoclimate and rapid environmental changes, he has played a crucial role in our numerous discussions on time and climate change. As a member of the advisory board for the project, Hall Bjørnstad has contributed friendly and insightful critique. A special thanks also to Susanne Leikam and Guro Flinterud, who were both involved in the research project at an early stage and contributed substantially to the research design.

We also thank the Department for Folklore and Ethnomusicology at Indiana University; the Department of History of Science and Ideas at Uppsala University; and the Department of Ethnology, History of Religions, and Gender Studies at Stockholm University for generously letting team members be visiting scholars during the writing of this book. Several of the research team members presented their research at an early stage at the roundtable 'The Future is Now: The Humanities and Environmental Crises' in Bloomington, Indiana, in February 2019. Our thanks to Hall Bjørnstad, Diane E. Goldstein, and Tabitha Rominger for organising this special event. We also thank the Department of Culture Studies and Oriental Languages at the University of Oslo and the Department of Cultural Anthropology and Ethnology at Uppsala University for facilitating manuscript workshops.

Thanks to the three anonymous referees who read our proposal and helped us clarify our ideas for this book. Finally, we thank Heidi Støa and Jordan Howie, who have been of great help proofreading the research proposal and the book manuscript, and the editorial team at Routledge for their helpful advice.

Oslo and Skallevold, 7 August 2020
Kyrre Kverndokk, Marit Ruge Bjærke, and Anne Eriksen

Introduction

1 Climate change temporalities

Narratives, genres, and tropes

Kyrre Kverndokk and Anne Eriksen

Climate change is intrinsically about time and temporality, but in a number of different ways. Climate change is about prospects for the future, about the relationship between present-day actions and future consequence, about comparing temperatures over long periods of time. Even more fundamentally, today's emission of greenhouse gases has long-lasting consequences on the Earth System. When atmospheric chemist Paul Crutzen and biologist Eugene Stoermer coined the term Anthropocene, claiming that humankind could no longer be overlooked as a significant geological agent, one of their main arguments was that anthropogenic emission of carbon dioxide (CO_2) will have climatic consequences over the next 50 000 years (Crutzen and Stoermer 2000, 17). Informed by such observations in science, recent theoretical debates on climate change temporality have been concerned with how climate change and the Anthropocene establish new conditions for historical understanding. Influential scholars, such as historians Dipesh Chakrabarty, Libby Robin, and Will Steffen, have pointed out that the long-term consequences of anthropogenic climate change and other human impacts on the Earth System require ways of thinking about history that take even planetary perspectives into account (e.g. Robin and Steffen 2007; Chakrabarty 2009; Robin 2013). In his seminal article of 2009 'The Climate of History: Four Theses', Chakrabarty argued that the history of humankind must be conceived as a history of the human *species*, and of how this particular species has influenced the development of the entire Earth System. A starting point for the discussion ignited by this article was the contention that the distinction between geological and human history had collapsed (Chakrabarty 2009).

Even the present book is part of this debate. We do not challenge Chakrabarty's contention regarding geological and human history. What we will question, however, is the utility of using very large and broad concepts of time, temporality, and history to understand climate change and its social and cultural implications. Our argument is that far more nuanced, detailed, and specific understandings and concepts are required to handle the challenges of a climatically changed world, politically and socially as well as scientifically. For this reason, the present book explores a variety of timescales and temporalities from narratives, experience, popular culture, and everyday life in addition to science and history – and the entanglements between them. The aim of doing so is not to contribute to new generalisations; rather, it is to argue that knowledge about the variety of temporalities and timescales that are involved is vital not only to understanding climate change but also to developing solutions to handle it and to summoning support for climate policies and actions.

Geological and historical time

The entwining of geological and historical timescales has long been at the centre of theoretical debates about climate change. However, these timescales are not as different as they might seem. Both operate at a macro level, as the history of the planet and the history of humanity. They are both global and chronological systems. Moreover, they are both organised in similar ways, as epochs and periods (and in geology also eras and eons), defined on the basis of specific characteristics of allegedly distinctive portions of time. Finally, both are academic constructs. What separates them are primarily the objects and processes they describe, and the duration of the time intervals within each system. Consequently, the insistence on their entanglement is not necessarily as theoretically radical as such a statement may appear. The notion of entangled timescales has nevertheless influenced historical understanding and work. On the one hand, the notion of the Anthropocene has challenged the historical disciplines; on the other hand, it has turned scientists into historians by inscribing human history into the history of the Earth. An example can be found in the introduction to the book *The Human Planet* by geographer Simon L. Lewis and Earth System scientist Mark A. Maslin:

> We humans are not just influencing the present. For the first time in Earth's 4,5 billion years history, a single species is increasingly dictating its future. In the past, meteorites, super-volcanoes and the slow tectonic movement of continents radically altered the climate of Earth and life-forms that populated it. Now there is a new force of nature changing Earth: *Homo sapiens*, the so-called "wise" people. (Lewis and Maslin 2018, 3)

This quote, and the book from where it is taken, does exactly what Chakrabarty suggested: the history of humankind is written as an entwined history of the human species and the development of the planet. Such an inscription of human history into a systemic, scientific frame facilitates a specific historical narrative of 'us' – the *Homo sapiens* – that is both progress-oriented and a history of decline, with the Anthropocene as the lamentable other side of the story (cf. Bonneuil and Fressoz 2017). The irony is that such a narrative has little or no room for describing and analysing agency, ideologies, politics, social structures, or cultural values. In other words, it has no room for specifically human experience nor for the perspective of the humanities. Consequently, the apparently radical theoretical perspectives of a new and entwined historiography of the Earth and our species can easily be criticised as being colonialist and imperialist, devouring human history instead of seeking to understand its deep and dramatic bearing on the planet.

We will argue that even if not incorrect, understanding the temporal aspect of climate change simply as the entwined scales of geological and historical time has limited value when it comes to investigating human experience, human agency, and human responsibility. It is not helpful for exploring how people imagine a climate-changed future or how they speak about it, and it does not provide analytical tools for understanding how the different temporal logics of science, politics, media, and everyday life intersect or collide in attempts to handle the climate crisis. Focussing on such particularities is first of all important for political and democratic reasons. People's experiences of and thoughts about climate change are important in their own right. The logics of politics and media are vital not only to general knowledge about climate change but also to its impact on real lives and life-worlds in real places. Second, diversified and specific knowledge about the temporal aspects that are involved is important because of the immensity of climate change issues and the challenges that humankind has created for itself and for the planet. Climate change is too large and too general to be left to large and general concepts. Finally, detailed, complex, and nuanced knowledge about climate change temporalities is theoretically significant. Even more than generalisations, sophisticated and precise terms, concepts, and perspectives are needed to understand the social and cultural aspects of climate change.

Rather than reflecting abstractly or philosophically on theories of temporality, history, and science, this book will investigate various timescales, timespans, intervals, rhythms, cycles, and changes in acceleration that are at play in climate change discourses. We will investigate how different kinds of text present different kinds of knowledge and understanding, distribute commitments and responsibilities, facilitate different kinds of action, and outline possible futures. Based on the tenet that 'climate change' is a concept that coordinates different phenomena, times, and spaces, we will explore how this coordination takes place and how it varies between texts and genres. Different genres imply different temporal logics, which also portray the relationship between past, present, and future in different ways. A climate-changed future is, for instance, about generational time and 'our children', it is about modelling sustainability on the example of traditional societies, about using concepts from the geological sciences to influence popular understandings of future climate change, and about revitalising traditional agriculture and handicraft as part of local climate activism. 'Climate change' is a concept that moves between science, politics, media, and everyday life, and thus between different discursive practices. In doing so, it creates meanings and implications of different kinds and invites actions and reactions that not only differ significantly from each other but also represent different temporal scales.

Climate change chronotopes

The point of departure for our investigations is the fact that 'climate change' is a conceptualisation of a highly abstract phenomenon. The Intergovernmental Panel on Climate Change (IPCC) defines climate change as 'a change in the state of the climate that can be identified (e.g., by using statistical tests) by changes in the mean and/or the variability of its properties and that persists for an extended period, typically decades or longer' (Matthews 2018, 554). This definition draws on a strictly scientific notion of climate as 'a statistical description in terms of the mean and variability of relevant quantities of certain variables (such as temperature, precipitation or wind) over a period of time ranging from months to thousands of years' (World Meteorological Organization, quoted in Hulme 2017, 2). Consequently, 'climate change' is characterised by not being directly observable. In that regard, it is characterised by 'nonlocality': it is everywhere, yet cannot be localised (Morton 2013). Moreover, the numbers, graphs, or texts by which climate change becomes a defined object are not just representations of an intangible phenomenon; to a certain extent, they *are* the phenomenon. This implies that to make 'climate change' a meaningful object for science, politics, public debate, and everyday life requires different linguistic and semiotic practices. To make it meaningful is not merely a matter of translating scientific knowledge about something into more broadly understandable language. The 'something' in itself – climate – needs to be conceptualised, expressed, narrated, and materialised to acquire substance and meaning. In vernacular and political settings, it may be an additional problem that the word 'change' normally carries highly positive values of gain, achievement, and action, rather than the sombre implications of a climatically changed planet.

This book examines a number of different entextualisations of the highly abstract concept of 'climate change', and a number of different discourses in which this happens. The terms chronotope and genre will serve us as important analytical tools. We understand genre as a reciprocal relation between form, function, and content. The form and functions of, for instance, scientific reports, popular science books, and news articles on climate change, obviously impact differently their (possibly shared) contents, also when it comes to the temporal dimension. Our understanding of genre implies that such differences are not due to just one single aspect but rather caused by the interplay between all three of them. It is this interplay that constitutes the logic of a genre, and is specific to it. We take inspiration from literary theorist Mikhail Bakhtin's argument that different narrative genres organise time and space by means of specific chronotopes. Literally meaning 'timespace', chronotope in Bakhtin's understanding refers to 'the intrinsic connectedness of temporal and spatial relationships' in a text (Bakhtin 1981, 85). We will use the terms genre and chronotope to identify how time and space are organised in a wide range of texts about climate change. By organising a genre temporally and spatially, the chronotope influences the content of a text.

Consequently, chronotopes also affect how human experiences, actions and possibilities might be portrayed (Ingemark 2016, 233). The presentation of actors, events, and phenomena are dependent on, though not determined by, the chronotope of the genre in which they appear. The chronotope is, however, not restricted to the generic level of a text. Chronotopes may even work on a micro level within a story, and make their mark on concepts or individual narrative motifs (cf. Ingemark 2016, 249). Because the chronotopic structure of a concept or motif is not necessarily identical to that of the genre in which it operates, chronotopic gaps and tensions may occur, and complex inter-chronotopic constructs may be created. On a macro level, on the other hand, the chronotopic construct of a text is also produced by intertextual relationships (Ingemark 2016, 234). While these are general aspects of chronotope theory, it is our contention that looking for such gaps, tensions, and inter-chronotopic constructs can offer important insights into different entextualisations of the abstract entity 'climate change'. What genres and chronotopes does this term invite? What kinds of narrative does the concept 'climate change' produce or at least make possible? In what ways do concepts, motifs, and genres interact or conflict when 'climate change' is the issue?

With the ambition to understand the chronotopic gap between the immediacy of the news genre and the long-term and large-scale concept of 'climate change', Chapter 5, by Kyrre Kverndokk, investigates how the Northern European heatwave of 2018 was depicted in news media. In the age of climate change, local weather news acquired an added global dimension. In this kind of news coverage, the longer term and larger scale of climate change is focalised into specific weather events. Relating local weather to climate change is, according to Kverndokk, 'not merely a matter of relating a particular case to statistic tendencies. It is also a matter of relating everyday life experiences to an abstract phenomenon made knowable through scientific expertise' (page 73).

As demonstrated by Lars Kaijser in Chapter 7, chronotopes may even relate to actual, physical space and include it in the narratives that are shaped by them. Kaijser investigates how the exhibitions at the grand public aquarium of Lisbon are constructed as narratives to be explored by the visitors that walk through them, but also how the visitors, by the use they make of the space, may produce other narratives with other chronotopes. To both the curators' intention and visitors' perception, however, being present in a physical space that is also a narrative is fundamental to the experience of the aquarium and its exhibitions. 'Walking the story' is the phrase that Kaijser makes use of to designate this double chronotopic function.

Camilla Asplund Ingemark argues in Chapter 4 that chronotopes contain traces of the different genres and spheres of communication they have been used in, and states that 'chronotopes do in fact have an agency in and of themselves – deriving from the multiple prior contexts in which they have been used' (page 49). Based on a corpus of answers to qualitative questionnaires, she demonstrates how vernacular notions of climate change are constructed by the means of a wide range of different narrative motifs, narrative tropes, and concepts. She states that 'climate change as a concept is capable of organising multiple temporalities, while at the same time being organised by certain other concepts of a similar nature' (page 49). She finds that 'crisis' and 'catastrophe' represent such concepts that both organise narrated climate change temporality and are given meaning by it.

Conceptual temporalities

While 'climate crisis' refers to a significant and decisive turning point, 'the climate catastrophe' and 'sustainable development' are both concepts that model possible future outcomes of the crisis. The terms are complementary, representing different, even opposite, stories. Even 'the climate crisis' is in this regard a plotted term or 'a point of time filled with significance, changing with a meaning derived from its relation to the end' (Kermode 1967, 47). One such end is the 'climate catastrophe'. The term is commonly used in several European languages, and it usually occurs in the definite singular, thus implying an understanding of a climate-changed future as one unified disaster that will affect us all. It is also in this sense, as an all-encompassing future, that the concept is used by politicians, environmentalists, and scientists when describing our prospects if efforts to dramatically reduce carbon emissions fail. The notion of 'sustainable development', including related terms such as the 'transition to a low-carbon economy', is a counterpart to the catastrophe. While the 'climate catastrophe' implies radical and dramatic change, 'sustainable development' represents a notion of seamless continuity from the present and into the future. As historian François Hartog has pointed out, the term indicates a future without any severe interruptions or revolutions (Hartog 2015, 200).

By means of these terms, the originally abstract notion of 'climate change' becomes narratively embedded. It starts to produce stories, or it is encapsulated by stories. This means not only that 'climate change' appears as more specific or concrete but also that its temporality and the events or processes that it refers to become parts of what can be called a narrative causality: things do not only happen chronologically, their temporal order is also given meaning as elements in causal chains that constitute specific plots, structured by a beginning followed by complications and leading to a solution, finally crowned with an ending or *coda* that often includes some kind of evaluative meta-commentary (Labov and Waletsky 1967).

The concepts of crisis, catastrophe, and sustainability are thus closely interrelated as well as narratively productive. Starting from the crisis, they form a bifurcated plot which in both versions (doom and perdition versus salvation and bliss) organises the relationship between the present and the future (Chapter 4; cf. Kverndokk 2017, 39–43). This narratively organised structure and its two main plots can be easily detected in political rhetoric and in policy documents. The chapters in this book also show it to be a common way of organising the temporal relationship between the present global situation and an imagined climate-changed future in both vernacular narratives and in public education (Chapters 3 and 7).

However, the book also explores other concepts and narratives that are related to, or derived from, this fundamental temporal structure. In Chapter 2, Diane E. Goldstein investigates how notions of 'immediacy' may throw light on narratives and utterances that are easily dismissed as ignorant or uninformed climate scepticism. Comparing immediacy as it is expressed in Donald Trump's populist rhetoric with that of informants who are worried about their jobs, the precarious economy, and their everyday struggles, she identifies a large complex of stories and narrative motifs that greatly impacts global climate discourse. She further demonstrates that, even though the way of actuating immediacy among the laypeople she quotes and Trump seems similar, the similarity is merely superficial. For the laypeople, the immediacy is often about handling a sense of powerlessness in everyday life. For Trump, on the other hand, the immediacy is about convincing his supporters and maintaining his position; it is a matter of holding onto power.

Marit Ruge Bjærke explores the notion of 'mass extinction' as it is used in international bestselling books in Chapter 8. The term situates the present loss of biodiversity into long-term global history, thus equating the eradication of species caused by human activity today with dramatic upheavals in deep time. Bjærke points out how the temporal aspects of the concept nonetheless change when it moves from the natural sciences, most notably geology, to other areas of discourse such as popular science and politics. Depending on the genres it enters, the notion of extinction takes on new meanings and different implications, and is productive of new narratives.

Tropes and tonality

Referring to the network or relationship between texts that generates related understanding in separate works, the term intertextuality was coined by philosopher Julia Kristeva (1980) and goes back to Bakhtin's notion of dialogism. The present book explores what can be called typological intertextuality, created by the means of specific tropes, tones, and meta-narratives (Baker 2006). In the various genres of climate change discourse, ranging from popular media to popular science and political rhetoric, 'the child' is a dominant trope. Normally set in plural form with a possessive pronoun, 'our' children or grandchildren indicate a notion of the future centred on the family. Just like the conceptual triad of 'crisis', 'catastrophe', and 'sustainability', the trope structures the relationship between the present and the future. The future becomes less abstract and more imaginable when transformed into the expectations and hopes that people have for their children and grandchildren. Moreover, it becomes 'our' responsibility in the present to secure these future expectations. When a climate-changed future is described in terms of 'the future of our children and grandchildren', it consequently is just as much about 'us' – the present-day adults and 'parents' – as it is about the children.

'Our children' and the notion of family time is used as an efficient way of establishing a convincing ethos in climate politics and environmentalism (Kverndokk 2020).

Lone Ree Milkær explores notions of temporality in what she has termed 'everyday-life climate activism' in Chapter 3, and she argues that this kind of activism is largely temporally organised by generational or family time. Family time, a term originally coined by social historian Tamara Hareven, 'relies on repetition and stability, exemplified by the expectance of knowledge transmission to children and grandchildren', Milkær writes (page 40; cf. Hareven 1977, 59–61). The activists that she has studied emphasise the importance of transmitting what they regard as sustainable everyday-life skills and practices to their children. This also brings them close to what can be called the meta-narrative of tradition, or the idea that certain kinds of knowledge 'should' or 'used to' be transmitted over generations, not only ensuring that useful skills and competences are passed on but also representing a kind of intragenerational principle of life.

The child-trope has also proven to offer a role to take on more directly. Greta Thunberg and the Fridays for Future movement draw heavily on the future-dimension embedded in this trope when they step forward as symbolic time travellers, travelling back in time from the future. Thunberg's oft-quoted words from her speech at the World Economic Forum in 2019 might serve as an example:

> Adults keep saying: 'We owe it to the young people to give them hope.' But I don't want your hope. I don't want you to be hopeful. I want you to panic. I want you to feel the fear I feel every day. And then I want you to act. I want you to act as if you would in a crisis. I want you to act as if our house is on fire. Because it is.
>
> (Thunberg 2019, 24)

The fire metaphor brings a disastrous climate-changed future into the present-day moment as an integrated part of the contemporary climate crisis. Thunberg's rhetoric is characterised by a circular temporal structure, or a loop. She places herself in the future, looking back at the present (Chapter 6; cf. Dupuy 2002). The noun *fear* indicates that this looped time travel is a deeply affective experience. She uses this affective experience to invoke the same kind of affect in her audience. Thus, there is something prophetic about her performance.

In Chapter 6, Isak Winkel Holm explores such kinds of prophetic experiences, focussing not on the prophet as a social role but on what he terms an aesthetic tone. 'The prophetic tone' is defined as 'the perception of the precatastrophic present [...] affectively charged by the imagination of a postcatastrophic future' (page 91). Holm distinguishes between the prophetic and the apocalyptic tone. The prophetic is 'intrahistoric and immanent in so far as it weaves together two moments in time, respectively a precatastrophic present and postcatastrophic future', while the apocalyptic 'is not *a loop in time* but rather a *leap out of time*' (page 103, Holm's italics). Holm explores the presence of the prophetic tonality in the television series *True Detective*. Through his analysis, he demonstrates how this tonality is also crucial for understanding the temporality of 'the climate catastrophe'.

As we point out in this book, several of the tropes and motifs that frequently occur in climate change discourses have a long cultural history. One frequently occurring trope is that of the 'simple man'. Historically, it has taken the form of, for instance, 'the savage', 'the peasant', 'the worker', or 'the people'. 'The simple man' represents, in one sense, authenticity, while at the same time being regarded as ignorant and unruly. Like 'the child', 'the simple man' is a trope that expresses asymmetrical power relations. It implies the need for authorities and experts to educate and govern. This notion of the ignorant man is identifiable in contemporary science as an 'imagined layperson' that in many cases works as a model audience for the dissemination of scientific knowledge (Maranta et al. 2003; cf. Chapter 2). This expert-layperson relationship has a long cultural history, which Anne Eriksen demonstrates in Chapter 9. Exploring a case of alleged climate deterioration due to kelp burning, she investigates how fishermen along the Norwegian coast were cast into the role of the simple man in late eighteenth-century texts. Conceived as authentic and innocent, but also skilled and competent in hard work in a harsh region, they came to embody local knowledge as well as local nature. Chapter 9 is also a study of the encounter between local and academic knowledge, with a micro-historical approach informing its understanding of '[t]he separation of (local) nature from knowledge about natural laws' and how this separation 'has made it possible to act upon nature in transformative and decisive ways, to build the modern world' (page 142).

This separation of nature from the scientific knowledge about it might be regarded as a topos in the grand narratives of modernity, in line with the separation of 'nature' and 'culture'. The terms 'nature' and 'culture' are two of the most used tropes in twenty-first-century cultural theory. With more or less explicit references to Bruno Latour or Donna Haraway, cultural theorists continue to insist that 'nature' and 'culture' today are entwined. The grand narrative of the division and re-entanglement of 'nature' and 'culture' is, for instance, the foundation for Chakrabarty's thesis that 'anthropogenic explanations of climate change spell the collapse of the age-old humanist distinction between natural history and human history' (Chakrabarty 2009, 201). Chakrabarty traces the distinction back to the early eighteenth-century philosopher Giambattista Vico and the so-called *verum factum* principle. In Chapter 10, John Ødemark goes back to Vico's texts and shows how the history of this divide is far more complex than it appears in Chakrabarty's article.

If, as Chakrabarty asserts, the Anthropocene calls for a new compact between natural and cultural history, a detailed account of how "we got there" in the past can perhaps also point forward to a new mode of convergence between these genres, states Ødemark (page 159).

Scales and dichotomies

While one of the aims of this book is to show that the temporal dimensions of climate change are more than an entanglement of the timescales of human and Earth history, another is to avoid the dichotomies that often come with the notion of scaling in climate change discourse. The abstract nature, global implications, and long-term perspectives of the IPCC definition and approach represents a stark contrast to human experiences and life-worlds. Derived from this, however, other dichotomies easily appear. When science is contrasted with lived experience, abstract knowledge is also opposed to the concrete and specific. Expert versus lay knowledge is another such contrast, in the same way as global versus local. The enumeration could go on, with words like long-term, analytical, disinterested, grand, theoretical, intellectual, powerful, political, detached, universal, fact-based, and international making up a cluster at one end of the axis, and short-term, embodied, partial, lived, vernacular, popular, bottom-up, concrete, local, social, embedded, and arbitrary lumped together at the other. On each side, the words tend to form fields or chains of equivalence. This implies, for instance, that vernacular knowledge *also* by implication comes to be considered as embodied, affective, local, short-term, partial, and so on and so forth, while expert knowledge *also* automatically is conceived of as universal, fact-based, disinterested, and so on. Moreover, the terms on each side come to work more or less as synonyms and stand-ins for each other.

Several problems spring from these false equivalences. One of them is that it is so easily forgotten that even experts are 'people', as Goldstein points out in Chapter 2, or that people may be skilled experts. Another is that the reason to engage with popular or vernacular knowledge (in all its alleged embedded and affective locality) frequently will be to educate, inform, correct, and enlighten. Even the insistence that the knowledge and experiences of 'ordinary people' can enrich scientific expertise by adding important new dimensions builds on similar dichotomies and carries with it an intrinsic risk of 'othering' lay knowledge. While averse neither to education nor to respecting other forms of insight than that of abstract science, our agenda through the chapters of this book does not align with either side of such a dichotomy. As our explorations will show, the genres, chronotopes, motifs, tropes, and tonalities of climate change discourse do not form a pattern of simple dichotomies. Quite to the contrary, they occur across any such imagined borders. They move, change, evolve, and migrate. Types of knowledge meet and get entangled; different temporalities or timespans do likewise. Personal, short-term, local, and embodied knowledge is integrated in powerful political speech and action, while scientific expertise and knowledge may be taken into use to bolster a personal and individual worldview. Only an empirical approach to the discourses and narratives of climate change and its intrinsic temporalities will produce insights into the wealth and variety of this discourse. It is our contention that such knowledge is necessary not just to understand what is going on, what is known, thought, felt, and experienced about climate change, but also to develop strategies and find approaches that may lead from 'climate crisis' to 'sustainability'.

References

Baker, Mona. 2006. *Translation and Conflict. A Narrative Account.* New York: Routledge.

Bakhtin, Mikhail M. 1981. *The Dialogical Imagination. Four Essays by M.M. Bakhtin.* Ed. Michael Holquist. Austin: University of Texas Press.

Bonneuil, Christophe, and Jean-Baptiste Fressoz. 2015. *The Shock of the Anthropocene: The Earth, History and Us.* London: Verso Books.

Chakrabarty, Dipesh 2009. 'The Climate of History: Four Theses'. *Critical Inquiry* 35 (2): 197–222.

Crutzen, Paul J., and Eugene F. Stoermer. 2000. 'The "Anthropocene"'. *The International Geosphere–Biosphere Programme (IGBP) Global Change Newsletter* 41: 17–18.

Dupuy, Jean-Pierre. 2002. *Pour un catastrophisme éclairé: quand l'impossible est certain.* Paris: Seuil.

Hareven, Tamara. 1977. 'Family Time and Historical Time'. *Daedalus* 106 (2): 57–70.

Hartog, François. 2015. *Regimes of Historicity. Presentism and Experience of Time.* New York: Columbia University Press.

Hulme, Mike. 2017. *Weathered: Cultures of Climate.* London: Sage Publications.

Ingemark, Camilla Asplund. 2016. 'The Chronotope of the Legend in Astrid Lindgren's Sunnanäng'. In *Genre – Text – Interpretation. Multidisciplinary Perspectives on Folklore and Beyond*, edited by Koski, Kaarina, Frog, and Ulla Savolainen, 232–250. Helsinki: Studia Fennica.

Kermode, Frank. 1967. *The Sense of an Ending: Studies in the Theory of Fiction.* Oxford & New York: Oxford University Press.

Kristeva, Julia. 1980. *Desire in Language. A Semiotic Approach to Literature and Art.* New York: Columbia University Press.

Kverndokk, Kyrre. 2017. 'Klimakrisens tid'. *Arr Idéhistorisk tidsskrift* 2: 33–47.

———. 2020. 'Talking about Your Generation: "Our Children" as a Trope in Climate Change Discourse'. *Ethnologia Europeae* 50 (1): 145–158.

Labov, William, and Joshua Waletsky. 1967. 'Oral Versions of Personal Experience'. In *Essays on the Verbal Arts*, edited by Helm, June, 12–44. Seattle: University of Washington Press.

Lewis, Simon L., and Mark A. Maslin. 2018. *The Human Planet How We Created the Anthropocene.* New Haven, CT: Yale University Press.

Maranta, Alessandro, Michael Guggenheim, Priska Gisler, and Christian Pohl. 2003. 'The Reality of Experts and the Imagined Lay Person'. *Acta Sociologica* 46 (2): 150–165.

Matthews, J.B. Robin, ed. 2018. 'Annex I: Glossary'. In *Global Warming of 1.5°C. An IPCC Special Report on the Impacts of Global Warming of 1.5°C above Pre-industrial Levels and Related Global Greenhouse Gas Emission Pathways, in the Context of Strengthening the Global Response to the Threat of Climate Change, Sustainable Development, and Efforts to Eradicate Poverty*, edited by Masson-Delmotte, Valerie et al., 539–562. Intergovernmental Panel on Climate Change.

Morton, Timothy. 2013. *Hyperobjects: Philosophy and Ecology after the End of the World.* Minneapolis: University of Minnesota Press.
Robin, Libby. 2013. 'Histories for Changing Times: Entering the Anthropocene?' *Australian Historical Studies* 44 (3): 329–340.
Robin, Libby, and Will Steffen. 2007. 'History for the Anthropocene'. *History Compass* 5 (5): 1694–1719.
Thunberg, Greta. 2019. *No One Is Too Small to Make a Difference.* London: Penguin Books.

Part 1

Vernacular notions of climate change temporality

2 'Where is global warming when you need it?'

The role of immediacy in vernacular constructions of climate change

Diane E. Goldstein

Introduction

On January 28, 2019, American President Donald J. Trump tweeted to his followers, 'In the beautiful Midwest, windchill temperatures are reaching minus 60 degrees, the coldest ever recorded. In coming days, expected to get even colder. People can't last outside even for minutes' (@real DonaldTrump). He continued, 'What the hell is going on with Global Wa[r]ming? Please come back fast, we need you!' Earlier in the same month he tweeted,

> Be careful and try staying in your house. Large parts of the Country are suffering from tremendous amounts of snow and near record setting cold. Amazing how big this system is. Wouldn't be bad to have a little of that good old fashioned Global Warming right now!
>
> (@realDonaldTrump, January 20, 2019)

The Trump administration's opposition to climate change action has been clear since shortly after his inauguration and even during the early days of his campaign, when he pronounced climate change to be a hoax created by China. His administration has rolled back environmental regulations, pulled the USA out of the Paris Climate Accord, removed environmental data from government websites, and pushed to alter Environmental Protection Agency climate change reports. While Trump's opposition to climate change policy reflects longstanding Republican opposition to government regulation in general and any policy specifically that would threaten the profits of fossil fuel companies, it also reveals Trump's confusion about climate change science. That confusion may be real or manufactured but it fits well with risk perception surrounding climate change and public struggles with concepts of temporal distance. Trump's articulated understanding of global warming is one that replaces the temporal distance of climate change with the immediacy of the everyday – a common focus of lay responses.

Lay understanding of climate change and global warming is challenged by complicated cause-and-effect relationships in which long-term climate patterns are harder to perceive than short-term localised weather.[1] Ironically, while a lack of response to the immediate crisis of climate change is of grave concern to scientists, a different kind of immediacy (in the sense of closeness, concurrence, proximity) pervades lay understandings. Over the last twenty years numerous international surveys of public understanding of climate change have argued that there is limited lay understanding of the scientific and technological issues involved, beyond a generalised knowledge of ecological deterioration and a general sense that humans are having a detrimental impact on the environment. Analysts frequently assert that an educated lay public is key to not only changing individual consumer behaviours but also providing public support for environmental regulations and initiatives. Lessons learned in the exploration of scientific assumptions about lay knowledge of medicine suggest, however, that experts frequently disregard the ability of laypeople to absorb information and therefore ignore lay patterns in thinking and interpretation which may be crucial in future policy endeavours. Analysis of lay perspectives identifies patterns in the role of immediacy (in space, time, and personal lifestyle) in vernacular understandings of climate change.

In what follows, this chapter will consider the shape and role of immediacy in lay constructions of climate change. The observations made here are based on a broad range of international reports and surveys of lay perspectives on climate change which include transcripts, ethnographic interview, and narrative. Analysis of this material identifies important patterns in the role of immediacy (in space, time, and personal lifestyle) in vernacular understandings of climate change. Two observations bear mention in relation to this material. Analysis of lay responses to climate change ebbs and flows with the collection of such data seeming to reach its high point in 2010, 2011, and 2012. The collection and presentation of that data does not appear to coincide with years that have focussed popular attention on climate concerns such as the issuing of Al Gore's film *An Inconvenient Truth*, the signing of the Paris Accord, the rise in popularity of Greta Thunberg, global protests, or fires, floods, and weather events that received significant coverage. While the urgency of climate change has accelerated significantly in the years since the production of some of these surveys of lay response, recent reports such as 'Communicating the Climate: From Knowing Change to Changing Knowledge' (Kleeman and Oomen 2019) and the Yale University report 'Climate Change in the American Mind: November 2019' (Leiserowitz et al. 2019) demonstrate that public concern, lay knowledge, and political action has not significantly increased in the years that have elapsed. Similarly, the nations represented in this material concentrate most heavily on Australia, the United Kingdom, Scandinavia, and the United States, indicating geographically based concern with public knowledge and behaviour.

Delayed destruction and the immediacy of now

Climate change experts note that climate change, as a crisis, lacks 'the immediacy of now' (Stern 2015). Environmental humanities scholar Rob Nixon calls this 'slow violence'. He writes:

> By slow violence I mean a violence that occurs gradually and out of sight, a violence of delayed destruction that is dispersed across time and space, an attritional violence that is typically not viewed as violence at all. Violence is customarily conceived as an event or action that is immediate in time, explosive and spectacular in space, and as erupting into instant sensational visibility. We need, I believe, to engage a different kind of violence, a violence that is neither spectacular nor instantaneous, but rather incremental and accretive, its calamitous repercussions playing out across a range of temporal scales. In so doing, we also need to engage the representational, narrative, and strategic challenges posed by the relative invisibility of slow violence.
>
> (2011, 2)

Nixon continues,

> Stories of toxic buildup, massing greenhouse gases, and accelerated species loss due to ravaged habitats are all cataclysmic, but they are scientifically convoluted cataclysms in which casualties are postponed, often for generations.... How can we turn the long emergencies of slow violence into stories dramatic enough to rouse public sentiment and warrant political intervention, these emergencies whose repercussions have given rise to some of the most critical challenges of our time.
>
> (2011, 3)

Because climate is not the weather and not the seasons, it is both hard to convey and easily seen as irrelevant or as a theoretical construct. Geographers Catherine Brace and Hilary Geoghegan argue that

> [t]here is a metaphysical and semiotic problem here with discussing in terms of a future date something that is made of the stuff of everyday life (for example weather) but which is not, in and of itself, *that stuff,* but aggregated, averaged, modified, smoothed, stripped of its outliers, rendered in statistical ways that remain mysterious to the majority.
>
> (2011, 291; emphasis in original)

The experience of climate change and its public understanding is distant not only in time but also in space. Once again, a quote from Donald Trump characterises the problem. Long aggrieved about changes in a product of personal import, Trump argued at numerous campaign rallies in 2016, 'You know, you're not allowed to [use] hair spray anymore because it affects the ozone'. Trump said, 'Hair spray's not like it used to be. It used to be real good' (Miller 2016). Complaining about modern regulation, Trump further characterises his puzzlement about the issue when addressing an audience of West Virginia coal miners:

> 'My hair look okay?' Trump asked the crowd. 'Got a little spray – give me a little spray. You know, you're not allowed to use hairspray anymore because if affects the ozone. You know that, right?' he said to laughter (Mackey 2016).

He continued,

> I said, "You mean to tell me" – 'cause you know hairspray's not like it used to be, it used to be real good,' he added, to more laughs. 'Give me a mirror. But no, in the old days, you put the hairspray on, it was good.
> (Mackey 2016)

Following those comments he noted,

> Today, you put the hairspray on, it's good for twelve minutes, right?' I said, "Wait a minute – so if I take hairspray and if I spray it in my apartment, which is all sealed, you're telling me that affects the ozone layer?" "Yes." I say, no way, folks. No way!
> (Mackey 2016)

Trump repeatedly suggests that he doesn't understand how the ozone destroying chlorofluorocarbons contained in aerosol propellants diffuse beyond an immediate (sealed) space. No matter how sealed Trump's apartment is and no matter how thick the walls are (he has suggested these act as a barrier), the gases mix, diffuse, and are transported into the stratosphere. Ozone depletion and climate change are not the same thing (although ozone is a greenhouse gas), but Trump uses the hairspray example repeatedly to comment by analogy on the folly of fossil fuel regulation.

Over the last few years Trump has declared that former US Vice President Al Gore should be stripped of his Nobel Prize, argued global warming is a hoax, and claimed it was created by and for the Chinese in order to change the US/China manufacturing picture. But it is Trump's assertion of time and space that interest us here, not his ongoing dismissal of climate science. Trump's muddled conflation of weather and global warming and his lack of understanding of spatial issues and fossil fuels, while perhaps more calculated than ignorant, reflect temporal and spatial challenges often present in lay understandings of climate change.

Yale University's climate survey program found in November 2018 that 74 percent of women and 70 percent of men believe climate change will harm future generations of humans, but just 48 and 42 percent, respectively, think it's harming them personally (Rogers 2018). In the 2019 Yale survey, only 38 percent of Americans reported thinking that people in the United States are being harmed by global warming 'right now' (Leiserowitz et al. 2019, 4). Making climate change *spatially* present is similarly a challenge. The Yale climate survey program found in 2019 that 44 percent of Americans think they will be harmed by global warming, but 64 percent think global warming will harm people in developing countries (Leiserowitz et al. 2019, 4). Such beliefs are underscored by news images of drought and famine as a result of heat waves in developing countries; the displacement of peoples uprooted by starvation, flood, and fires; the spread of diseases such as dengue fever, zika, malaria, chikungunya, and cholera; sea level rise, glacial retreat, and landslides; and changes in animal, bird, and plant ecosystems – all seemingly far away, affecting other people and other places – spatial and temporal immediacies clearly not understood in the same way by the scientific community and the lay public.

Lay and expert knowledge

It is perhaps helpful in trying to understand public perception and construction of climate change to think about the analogy of how laypeople make sense of health and illness and how the discourses of medical science respond to lay belief and knowledge. It is increasingly an accepted factual claim that if public health research is to be at all effective it must incorporate, develop, and respond to the theoretical and conceptual insights offered by laypeople. Crucial to the incorporation of that information is a degree of respect for lay explanatory models, a recognition of the complexity of lay perspectives, and an understanding that narrative and lay discourse have embedded within them explanations for what people believe and why – explanations which ultimately determine social action. This requires a reconceptualisation of lay knowledge as both accessible and useful.

Perceptions on the part of people in the medical field of the use of health information by members of the lay community are products of what sociologists Maranta et al. (2003) refer to as the 'imagined lay person'. They note:

> [...] between expert and lay persons there is a division of labour that is based on epistemic asymmetry: the experts are supposed to be knowledgeable and the lay persons are ignorant [...] Within this context, 'imagined lay persons' are conceptions of lay persons as they are manifested in the products and actions of the experts [...].
>
> (2003, 151)

Maranta et al. continue, 'Imagined lay persons need not be explicit. Nor need they have any resemblance with real lay persons. Rather imagined lay persons are functional constructs in expertise' (2003, 151). The notion of the imagined layperson is based on a deficit model, in which the lay population is understood by experts as *tabula rasa,* ignorant of the intellectual content, research methods, and organisational forms of science. Despite the fact that in the real world, those who might be considered laypersons are extremely diverse in terms of a full range of training, competencies, and capabilities (some may be trained engineers, technicians, or science librarians), the group of imagined laypersons is taken by experts to be an undifferentiated aggregate or collective, jointly lacking in scientific knowledge and constrained in both ability and action. As Maranta et al. (2003) argue, experts perceive lay users as uncritical readers unable to assess the credibility, authority, relevance, timeliness, or motivations of health information. The same is true of expert attitudes towards lay understandings of climate change. Social psychologist Glynis M. Breakwell (2010), for example, interprets one study of public risk perception by noting that '[t]he lay public did not have a holistic risk representation that was *capable* of relating causes to consequences of climate change' (Breakwell 2010, 859, emphasis added).

Moving beyond the *tabula rasa* theory of lay knowledge, psychiatrist and anthropologist Arthur Kleinman proposed a theory of explanatory models which explored the vastly different notions of health and disease held by individuals and groups. Explanatory models are 'notions about an episode of sickness and its treatment that are employed by those engaged in the clinical process' (Kleinman 1981, 105). The explanatory model held by patients influences receptivity to health promotion messages and choices of health behaviours. Characteristics of patient explanatory models were (1) specific to each episode of illness; (2) calibrated to practical realities and pressing problems; and (3) focussed on one's experience, frustrations, sources of knowledge, and resources at hand (Kleinman 1980, 105). Kleinman's points highlight lay health beliefs as being both situated and oriented towards the practical. The importance of both the situated and the practical appear in contrasting expert and lay definitions of what constitutes health and illness, reportable symptoms, and incapacity. In the Western biomedical paradigm, health is seen as the absence of disease and the appropriate functioning of biological and psychophysiological processes in the individual. Lay concepts of health often focus more practically on ability to function or on social well-being evidenced by physical fitness, energy, vitality, absence of pain, and the ability to meet work expectations and maintain social relationships (Calnan 1987).

Like health and medicine, climate change science is a challenge to public comprehension. The science is complex, global economics are intertwined, policymakers purposefully or accidentally confuse matters, and scientists themselves hotly debate both causes and effects. Nevertheless, lay comprehension is critical to climate change policy, and therefore understanding lay perspectives is central to the work of tackling the climate crisis. As Brace and Geoghegan note, climate change can be seen 'as a relational phenomenon', 'understood on a local level, with distinctive spatialities and temporalities'. It is 'felt, sensed, apprehended emotionally as part of the fabric of everyday life in which acceptance, denial, resignation and action co-exist as personal and social responses to the local manifestations of a global problem' (Brace and Geoghegan 2011, 284). Like lay concepts of health, the perception of the problem focusses on it as a practical and situated one – concerned with the immediate, that which is present in both spatial and temporal terms. While not the same as expert knowledge on climate change, lay knowledge engages with the local and the present moment in ways that can be productive if heard and taken seriously.

Time

Climate change researchers Mike Hulme, Suraje Dessai, Irene Lorenzoni, and Donald R. Nelson argue that scientific narratives about climate change are based on intervals of decades to centuries while most people make decisions and structure their behaviour on more immediate timescales (Hulme et al. 2009, 201). Lydia, when interviewed, illustrated the lengthening of time frames dampening immediacy by pointing to the distance down the road of the threat. She said,

> Don't talk so much about 2050, they will think 'Well I am going to be dead by then,' so they don't take much notice, or '2050 is a long way away and I will see about it, you know, at 2045', rather than knowing that it is a continuous process.
>
> (Hanson-Easey et al. 2015, 228)

Brett, a factory foreman from Australia in 2015, contrasted the urgency of climate change with the perceived immediacy of financial need. He commented on the Australian carbon tax by saying,

> Well the government are bloody buggering around with that with climate change if you like by the tax. And the tax – if you look at climate change and then what the government is doing with the gas tax, the carbon tax as 'what does it do to put prices up; what does it do for me to get an offset for prices going up; what is it doing with the job market'.

He continues,

> I don't think there is a lot of people – I mean I am sort of adlibbing here a bit – that worry about: 'Oh bugger me, in forty or fifty or a hundred years' time, it is going to wreck the coasts and the rivers and the climate'. So they relate it back to jobs, money, lifestyle [...].
>
> (Hanson-Easey et al. 2015, 231)

Brett also focussed his comments on the immediacy of time-consuming life issues and exhaustion in contrast to time to taking action. He noted,

> And then you get into a spiral and you think, 'This is pretty serious, I will do something about this. I will think about it'. But what they are

doing is they are thinking about, 'Jeez I'm bloody tired. I just worked a twelve-hour shift'.

He follows this by saying,

> I watched bloody half an hour of something where I don't have to think about anything and just some wacker [idiot] on television. Then it is the next morning. Get the kids to school. Get the kids to football. So they are in survival mode [...].
>
> (Hanson-Easey et al. 2015, 230)

Cheryl, Victoria, and Xavier agreed. Speaking together they said:

CHERYL: Oh, that is down the track. It's not affecting me right now.
VICTORIA: And you can see why – there are so many more important [...]
XAVIER: I have to get a job – pay for the mortgage.

(Hanson-Easey et al. 2015, 227)

Geographers and environmental researchers Johanna Wolf and Susanne C. Moser note that public interest in the reality of climate change is dependent on concurrent events, such as weather extremes, or non-climatic events, like terrorist threats, economic recessions, or major public controversies (Wolf and Moser 2011, 548). In other words, the immediacy of climate change is gauged with reference to the immediacy of other threats to survival.

The construction of climate change immediacy also denies the idea of a continuous timeline by some lay respondents emphasising the cyclical nature of climate events. As anthropologist Linda Connor and epidemiologist Nick Higginbotham note, cycles situate humans as intertwined with nature, not controlling or affecting natural process (2013). One woman noted,

> After the '67 bushfires raged through ... there were years of drought and years of decent rainfall. We've just had probably about three years of decent rainfall and we're probably going to have another two years of decent rainfall and probably four or five years of drought. To me [climate change] it's cyclic.
>
> (Denniss and Davison 2015, 146)

A 60-year-old woman said in 2013, 'I would like more proof as to how much climate change is happening through man- made means or whether it is just a natural cycle of events' (Connor and Higginbotham 2013). A retired man from Australia interviewed at the same time agreed. He said,

> I do not believe in climate change. It is a natural cycle. I remember as a child the weather was hotter – ten days in a row of 100 degrees plus. It is a conspiracy by government to introduce a new tax.
>
> (Connor and Higginbotham 2013)

A 45-year-old Australian technician was more specific, citing evidence for the cyclical nature of climate events. He said, 'I have lived in the area for 48 years and the sea level is the same. There seems to be a flood following a low depression every 10–15 years' (Connor and Higginbotham 2013).

Some participants focussed attention away from immediacy by pointing to past behaviour to explain the current state of affairs. Several American respondents blamed climate change on atomic bomb testing, space travel, or other forms of older and often governmental rather than contemporaneous environmental abuses. Wilber indicated,

> I've always felt that when they had that bomb it had an awful bearing on the change of our weather. The A-bomb. They had those tests [...] Just seemed like here things have changed ever since, it's become more torrent, the weather here in the past few years [...]

Asked what that means, he replied, 'Violent, violent, yeah. The weather is very changeable. They say that didn't have nothing to do with it, but I still feel that it did somewhere along the line' (Kempton 1991, 191; italics in original). Susan said,

> I have my own private theory [pause] *What's that?* That every time they shoot something up in space it disturbs things up there! *There could be something to that.* I've been told that I have no foundation for that, but it just seems every time something happens we get this strange type of weather [...] *Like what* [...] Well for instance, tornadoes were very rare in this section of the country [...] tornadoes and violent storms [...] it used to be rather calm here.
>
> (Kempton 1991, 191; italics in original)

These comments on prior human encroachments on nature address the immediacy of now by lengthening the timeline of effect. The notions of encroachments on time and money, the construction of natural cycles, and backward and outward looking notions of blame, point to time as a discursive site imbued with localised, contextualised, and culturally salient understandings of climate change.

Space

The impact of perceptions of temporal immediacy in lay perceptions of climate change is mirrored by a similar significance of perceptions of *spatial* immediacy. Most people experience some form of what social psychologists call *optimistic bias* in their lives, in which people overestimate the likelihood of positive events happening to them and underestimate the likelihood of negative events. Optimistic bias has been found in association with a significant number of environmental issues. A large, eighteen-nation survey found, for example, that individuals globally believe that across a number of environmental issues they are safer than others living elsewhere and that they are safer than future generations (Gifford et al. 2009). In other words, respondents to the survey demonstrated both a temporal and a spatial bias. Wolf and Moser argue that

> climate change is as yet perceived by most people in developed countries as a distant threat that is removed from their lives both spatially and temporally. More specifically, climate change risks are perceived as non-personal, concerning the future, other places and people, and other species (plants and animals).
>
> (2011, 548)

Notions of cognitive mapping might help in understanding our spatial optimistic bias. Cognitive maps are made up of the knowledge and internal representation of the structures, entities, and relations of space – in other words, the internalised reflection and reconstruction of space, in thought (Kitchin 1994, 1). They include the social, cultural, and individual awareness; perception; information; images; and beliefs about our environment. Individual cognitive maps of climate change point out the importance of personal and cultural experience in one's view of the world. As Brace and Geoghegan note,

> [c]limate change can be observed in relation to landscape but also felt, sensed, apprehended emotionally as part of the fabric of everyday life in which acceptance, denial, resignation and action co-exist as personal and social responses to the local manifestations of a global problem.
>
> (2011, 1)

Optimistic bias says not here, not me, others live in more dangerous places; 'I have lived here all my life, it never floods here, wildfires never happen here'. The Yale climate change survey found that a moderate percentage of their survey respondents held the perception that climate change is a danger mainly to geographically and temporally distant people, places, and non-human nature (Leiserowitz 2009, 12).

Interviews with laypeople indicate that discussions of temperature variation are understood in terms of their personal place of habitation, gauged upon the basis of local daily swings of 3 to 9 degrees Fahrenheit rather than in terms of far-flung global geophysical and ecosystem effects (Kempton 1991). Individuals interviewed conceived of temperature change in terms of their own local understanding of hot weather rather than the more distant, but also more profound, environmental consequences. Asked, '*if the weather gets a lot warmer, do you think it would be good, bad or neutral*', Paige remarked,

> 'I think it would be bad; I think it would be terrible. *Why?* Well, I think people react differently in warm weather than when it's cooler. I think it

> has an effect on attitudes – behaviour … I mean in the prison system especially, where the people are just, you know, stuck in there, and they've got to let off steam. So, sure.

She continues,

> *So you think in prison it makes people more violent?* Sure, but outside the prisons, too, cause I even see it at work; you know, when the weather is extremely warm, people tend to be, you know, a little hot tempered. I think you know, their blood boils.

Paige finished, 'And when the blood boils in the body, it goes to the head, and next thing you know, there's, you know, an explosion […] I've seen them react that way' (Kempton 1991, 189; italics in original). Paige understood warming only through her own immediate place of habitation or those who live in a similar setting and was concerned mainly about the emotional impact of experiencing extreme heat. Locality and proximity were a core part of her model of climate change immediacy. Melting of glaciers, sea level rise, species extinction, impact on food supply, and so on, did not figure into her perception of global warming. Others also stressed the effects of local environmental conditions on perceptions of climate change, focussing on a lack of climatic danger in their immediate surround.

Cindy felt a lack of spatial immediacy until it ended up on her doorstep. She said:

> The thing that really got me was when, I think it was two years ago, we stayed down at the shore for a week and the last day we were there, they closed the beaches and I thought, 'Boy if you can't even swim in this water, it's got to be really bad'.
>
> (Kempton 1991, 194)

It is the case, however, that climate changes in one's own locale affected one's acknowledgement of global significance. A Welsh study of climate change public perception found that a quarter of survey respondents reported having been directly affected by flooding, with one in twenty people having experienced property damage. The report goes on to note that those who have experienced flooding are more likely to see their local area as vulnerable to climate change and are therefore more accepting of the overall global concept of climate change (Capstick, Pidgeon, and Whitehead 2013, 27). Psychologists note that when an object is perceived to be psychologically close to the self, it tends to be perceived in concrete terms, whereas objects perceived as distant from the self are construed more abstractly. Thus, climate change that is perceived as close will be viewed more concretely and be more likely to provoke action than that which is viewed as distant from the self (McDonald et al. 2015, 110). Farmers, for example, who experience drought or floods are more likely to express concern about climate change than those who have not had such experiences (Prokopy et al. 2015).

Understanding spatial immediacy is, however, complicated by issues of the location of responsibility and power to create change. John, who understood the local and global threat, focussed on those in authority, spatially removing himself from the immediacy of responsibility or ownership. He said, 'Well, environment, there's nothing you can do about it. I'm not going to upset myself over it. That's all. I'm not in no position. I have no authority' (Kempton 1991, 194).

Then he returns to not only spatial distance measured by power but also spatial distance used to create blame measured geographically. Another gentleman named John allows for the global picture of climate change, but uses the global picture to identify foreign responsibility. He said,

> And the other thing is that it always comes back when – and I follow on exactly what you have said, is they say: well what we do – how is that going to affect when you have got China building more bloody coal fire plants. We are still shipping coal. We are doing all of this. Why – how is this possibly going to make any difference other than to my pocket now.
>
> (Hanson-Easey et al. 2015, 225)

Most of these responses involve embodied knowledge, emphasising, as Brace and Geoghegan call it, 'the immediacy of landscape in an emergent subjectivity' (2011, 293). Spatial immediacy focusses on our experience of climate change in the places that we experience directly or that we have had reason to locate on our cognitive maps. Feelings of distance from authority, responsibility, opportunity, or power mediate personal and social responses to spatial immediacy, placing global and even local understanding and action just on the periphery of the recognised landscape.

Like John, Andrew explains that he does not consider climate change in his voting or political affiliations. He prioritises 'more immediate concerns' that he can get 'emotionally involved' with rather than the abstractions of climate change. Andrew recognises that nature is seen as an 'endless resource that can be exploited for the benefit of humans', but he deflects personal responsibility: 'I am so preoccupied with my own issues that I don't have this ambition of saving the world' (Denniss and Davison 2015, 149).

As John Fiske notes, localising knowledges

> function not to extend a great vision over the world, but to produce a localised social, ethnic, communal sense of identity. They create cultures of practice, ones that develop ways of living in the world and which seek to control only those ways of living rather than the world in which they live.
>
> (1993, 19)

Lay knowledges and climate change

Like characteristics of lay explanatory models of health, lay models of climate change are calibrated to practical realities and pressing problems, and focussed on one's experience, frustrations, sources of knowledge, and resources. Lay knowledge is situated and practical, tied to immediacy both temporally and spatially. Just as non-biomedical or lay concepts of health often focus on ability to function, or on social or spiritual well-being evidenced by physical fitness, energy, vitality, absence of pain, feeling healthy, and the ability to maintain social relationships (Calnan 1987), lay concepts of climate change focus on the now and the proximate, the nature of experience, trust in personal observations, and the nature of essential necessities. Sometimes that immediacy leads one to problematic constructions such as the conflation of weather with climate, but at other times it prioritises cultural concerns and needs that are hard to argue with (such as the deeply urgent temporal and fiscal needs of the economically disadvantaged, noted by Brett above).

Just as experts imagine laypeople in a stereotypical way, highlighting their lack of knowledge and their inability to think and read critically, laypeople imagine experts as unaware of or unresponsive to the real nature of immediate needs. But immediacy poses a problem for experts as well. Biologist Jim Jarvie, director of the climate, environment, and energy unit at Mercy Corps, said in speaking about fellow experts,

> I've found that in conflict areas, if you raise the visibility of climate change – which we did in Afghanistan – we can just see their eyes rolling up, saying, 'Oh good Lord', because it's not the immediacy of *now*. Indeed, adapting to climate change may not be as pressing a concern for the world as stopping bullets or Islamic State.
>
> (Stern 2015, 1)

He continued: 'Yet failing to tackle climate pressures now will carry severe future consequences for both sustainable development and security, those working on building peace in fragile states' (Stern 2015, 1). One of the often-overlooked issues tied to lay and expert knowledge is the indisputable fact that experts are themselves people, steeped to a degree in the spontaneous human responses that comprise lay knowledge. Those responses do not easily move beyond what psychologist Abraham H. Maslow (1970) and sociologist Kari Marie Norgaard (2011) refer to as the 'hierarchy of needs', which places immediate needs and constructions ahead of those perceived to be long-term or distant. As Norgaard notes, 'people cannot think about climate change', in part 'because they are too consumed with solving the problems of the present' (Norgaard 2011, 75).

The quotes from Donald Trump which open this chapter are a case in point, placing immediate needs and constructions ahead of concerns perceived to be more distant. Trump's penchant for communication using non-scripted speeches and tweets demonstrates his rhetorical engagement with the immediate. But lay immediacy is different. For Trump immediacy is tied to the process, goals, and ends of exerting power, while lay immediacy is so often focussed on powerlessness.

Involving laypeople more intensely in climate change will require efforts to address the different perceptions of temporal and spatial immediacy at the core of lay (and some expert) conceptions of the problem and its solutions. Understanding public engagement or lack of engagement in climate change requires abandoning a deficit model of lay knowledge and replacing it with a model that recognises lay concerns as situated and practical.

Note

1 Similar to Trump's comments about global warming, Kari Marie Norgaard reports a story of a woman in western Norway who gets on a bus with her groceries. She writes: 'Despite the fact that it is mid-January, there is no snow on the ground. As she paid her fare, she remarked to the driver, "This global warming is a good thing!"' (2011, 98).

References

Brace, Catherine, and Hilary Geoghegan. 2011. 'Human Geographies of Climate Change: Landscape, Temporality, and Lay Knowledges'. *Progress in Human Geography* 35 (3): 284–302.

Breakwell, Glynis M. 2010. 'Models of Risk Construction: Some Applications to Climate Change'. *Wiley Interdisciplinary Reviews: Climate Change* 1 (6): 857–870.

Calnan, Michael. 1987. *Health and Illness: The Lay Perspectives*. London: Tavistock Publications.

Capstick, Stuart B., Nicholas F. Pidgeon, and Mark S. Whitehead. 2013. *Public Perceptions of Climate Change in Wales: Summary Findings of a Survey of the Welsh Public Conducted during November and December 2012*. Cardiff: Climate Change Consortium of Wales.

Connor, Linda H., and Nick Higginbotham. 2013. '"Natural Cycles" in Lay Understandings of Climate Change'. *Global Environmental Change* 23 (6): 1852–1861.

Denniss, Rebecca Joy, and Aidan Davison. 2015. 'Self and World in Lay Interpretations of Climate Change'. *International Journal of Climate Change Strategies and Management* 7 (2): 140–153.

Fiske, John. 1993. *Power Plays, Power Works*. London: Verso.

Gifford, Robert, Leila Scannell, Christine Kormos, Lidia Smolova, Anders Biel, Stefan Boncu, Victor Corral et al. 2009. 'Temporal Pessimism and Spatial Optimism in Environmental Assessments: An 18-Nation Study'. *Journal of Environmental Psychology* 29 (1): 1–12.

Hanson-Easey, Scott, Susan Williams, Alana Hansen, Kathryn Fogarty, and Peng Bi. 2015. 'Speaking of Climate Change: A Discursive Analysis of Lay Understandings'. *Science Communication* 37 (2): 217–239.

Hulme, Mike, Suraje Dessai, Irene Lorenzoni, and Donald R. Nelson. 2009. 'Unstable Climates: Exploring the Statistical and Social Constructions of "Normal" Climate'. *Geoforum* 40 (2): 197–206.

Kempton, Willett. 1991. 'Lay Perspectives on Global Climate Change'. *Global Environmental Change* 1 (3): 183–208.

Kitchin, Robert M. 1994. 'Cognitive Maps: What Are They and Why Study Them?' *Journal of Environmental Psychology* 14 (1): 1–19.

Kleeman, Katrin, and Jeroen Oomen, eds. 2019. 'Communicating the Climate: From Knowing Change to Changing Knowledge'. *RCC Perspectives: Transformations in Environment and Society*, no. 4.

Kleinman, Arthur. 1981. *Patients and Healers in the Context of Culture: An Exploration of the Borderland between Anthropology, Medicine, and Psychiatry.* Berkeley: University of California Press.

Leiserowitz, Anthony. 2009. 'International Public Opinion, Perception, and Understanding of Global Climate Change'. Report for Yale Program on Climate Change Communication.

Leiserowitz, Anthony, Edward Maibach, Seth Rosenthal, John Kotcher, Parrish Bergquist, Matthew Ballew, Matthew Goldberg, and Abel Gustafson. 2019. 'Climate Change in the American Mind: November 2019'. Joint report for Yale Program on Climate Change Communication and George Mason University Center for Climate Change Communication.

Mackey, Robert. 2016. 'Donald Trump's Hairspray Woes Inspire Climate Denial Riff'. *The Intercept*, May 6, 2016. https://theintercept.com/2016/05/06/donald-trumps-got-hairspray-riff-hes-gonna-use/.

Maranta, Alessandro, Michael Guggenheim, Priska Gisler, and Christian Pohl. 2003. 'The Reality of Experts and the Imagined Lay Person'. *Acta Sociologica* 46 (2): 150–165.

Maslow, Abraham H. 1970. *Motivation and Personality.* New York: Harper and Row.

McDonald, Rachel I., Hui Yi Chai, and Ben R. Newell. 2015. 'Personal Experience and the 'Psychological Distance' of Climate Change: An Integrative Review'. *Journal of Environmental Psychology* 44: 109–118.

Miller, Sara G. 2016. 'Hair Spray vs Ozone? Trump Makes Outdated Complaint'. *Live Science*, May 6, 2016. https://www.livescience.com/54677-trump-hairspray-ozone-layer.html.

Nixon, Rob. 2011. *Slow Violence and the Environmentalism of the Poor.* Cambridge, MA: Harvard University Press.

Norgaard, Kari Marie. 2011. *Living in Denial: Climate Change, Emotions, and Everyday Life.* Cambridge: MIT Press.

Prokopy, Linda S., J. Gordon Arbuckle, Andrew P. Barnes, V. R. Haden, Anthony Hogan, Meredith T. Niles, and John Tyndall. 2015. 'Farmers and Climate Change: A Cross-National Comparison of Beliefs and Risk Perceptions in High-Income Countries'. *Environmental Management* 56 (2): 492–504.

Rogers, Adam. 2018, 'The Climate Apocalypse Is Now and It's Happening to You'. *Wired*, November 28, 2018. https://www.wired.com/story/the-climate-apocalypse-is-now-and-its-happening-to-you/.

Stern, Rachael. 2015. 'Climate Change Lacks 'The Immediacy of Now' in Conflict Zones – Experts'. *Thompson Reuters Foundation News*, October 23, 2015. http://news.trust.org//item/20151023090515-4di1h.

Wolf, Johanna, and Susanne C. Moser. 2011. 'Individual Understandings, Perceptions, and Engagement with Climate Change: Insights from in-Depth Studies across the World'. *Wiley Interdisciplinary Reviews: Climate Change* 2 (4): 547–569.

3 The great re-skilling

Understandings of generation, tradition, and nostalgia in everyday-life climate activism

Lone Ree Milkær

Introduction

In recent years, movements such as Fridays for Future and Extinction Rebellion have established themselves on the scene of global climate activism. Yet, the global trends of climate activism do not necessarily imply demonstrations and people shouting slogans, as some activists choose to act in ways that relate more closely to their everyday lives (Barr and Gilg 2006, 910; Dowling 2010, 489; O'Brien, Selboe, and Hayward 2018). The everyday-life activists focus on the importance of recycling their waste, reducing consumption and travel, and planting vegetables or bee-friendly flowers in their gardens. These activities can be seen as ways of reacting to and coping with the abstract notion of climate change. Climate change is of course a scientific concept at its core, but it becomes much more when it is confronted and operationalised in everyday life.

To the activists, it is evident that climate change will have an impact on future living conditions all over the world and that climate change mitigation will include a reorganisation of society on a structural, political, and everyday level. The abstract concept of climate change can be interpreted as a temporal process which connects the future climate changes to a present feeling of urgency and crisis for the activists. An awareness of the future amongst climate change activists is perhaps evident, but nonetheless important to an understanding of the notions of time connected to climate change activism. It connects the notion of climate change to the abstract temporality of the scale of past, present, and future, which is what the following text will elaborate.

This chapter is a case study of grassroots, everyday-life climate change activism that examines how climate change mitigation is performed in everyday practices among local climate activists in the organisation Sustainable Lives.[1] I ask how the notion of climate change is incorporated in vernacular ways of connecting past, present, and future, and I answer the question through analysing the concept of *gjenkunning*. Gjenkunning is a word in Norwegian which means something along the lines of 're-skilling' but does not translate directly into English (as shall be seen below). The analysis is primarily based on texts written by activists to introduce and promote the organisation Sustainable Lives.[2]

Tradition, generation, and nostalgia

The concepts central to the following analysis are tradition, generation, and nostalgia. In different ways, these concepts create notions of continuity and connection as well as a relationship to place. All of these concepts connect time to perceptions of place in that they situate time by connecting it to some specific place and to some person or group of people. Traditions do not exist if no one upholds them, generations are people who live in different times, and nostalgia is the longing for a home, literal or metaphorical, in another time. These three concepts will assist in establishing connections between vernacular understandings of time and space and the notion of climate change as seen in everyday-life activism of the activists in Sustainable Lives. I will elaborate on the perspectives used in the analysis in this short introduction and try to untangle the otherwise entangled concepts of tradition, generation, and nostalgia for this purpose.

Folklorist Henry Glassie begins his seminal article from 1995 on tradition with a plea to the reader to 'Accept, to begin, that tradition is the creation of the future out of the past' (Glassie 1995, 395). I accept this premise and examine how this creation of the future out of the past takes place in a present. Any tradition must carry some sort of meaning for its users and this makes the meaning of tradition inseparable from the present and from a present interpretation of the content and values connected to it (Bauman 2004, 147). Tradition is in broad terms a concept which establishes continuity between past, present, and future and presumes this building of continuity as an essential value: 'In its common sense meaning, tradition refers to an inherited body of customs and beliefs' (Handler and Linnekin 1984, 273). The use of the word 'inherited' implies some kind of transmission and connection of elements and that these elements be situated in a long chain of similar and stable elements that are passed on unchanged through time. It is assumed that traditions have existed in the past and will exist in the future. The concept of tradition is based on this very assumption (Eriksen and Stensvold 2002, 84) and when people ascribe these characteristics of tradition – continuity, stability, inheritance – to everyday phenomena, it is a way of placing objects, cultural elements, processes, or beliefs into a temporal continuity. The process of ascribing tradition to everyday phenomena as a characteristic can be named traditionalisation (Eriksen and Stensvold 2002, 86–87). I will apply the concepts of tradition and especially traditionalisation as an analytical frame to understand how the everyday-life activists of Sustainable Lives relate to temporal continuity through their activities.

A slightly different concept used to describe perceptions of the past is nostalgia. The concept of nostalgia originated in the seventeenth century as a psychiatric diagnosis describing pathological homesickness; the term combines the Greek *nostos* for 'homecoming' and *algos* for 'pain' (cf. Johannisson 2001, 17; Boym 2001, 3–4). Originally nostalgia described a very specific yearning for a childhood home or the landscapes of a home region, i.e. the

Alps. In the twentieth century, the meaning of the concept shifted from primarily describing a longing for a specific place to describing a longing for a specific time, a 'before' (Johannisson 2001, 8; Cashman 2006, 139; Wilson 2015, 479; Smith and Campbell 2017, 614). This 'before' is closely connected to place, as nostalgia now designates a longing for the comfort and homeliness of childhood, and thus in the broader sense a longing for a place where you used to belong. Zygmunt Bauman describes this view of the past as a 'building site for comfort zones' (Bauman 2017, 65). This building site works as a starting point to an understanding of nostalgia as a longing for an idealised 'before' connected to the feeling of home and belonging, which combines elements of time and space (Johannisson 2001, 32). I use the concept of nostalgia to analyse gjenkunning as a synthesis of tradition and generation.

The texts analysed in this chapter connect both nostalgia and tradition to an idea of transmission through generations, as shall be discussed below. Generation can be seen as the organising principle of tradition and nostalgia, the anchoring of the perceived past in specific times and specific places, through associations with, for example, grandparents, children, and neighbours, and in specific localities such as a neighbourhood or the countryside. To connect the concept of generation to the notions of time, I will use the theoretical frame of family time, inspired by historian Tamara Hareven, as used by folklorist Kyrre Kverndokk: 'Family time is [...] a notion of experienceable time in between the individual lifespan and historical time' (Kverndokk 2020, 149). In this chapter, I ask how family or generation relate to the abstract time of climate change and how this further connects to the practice of everyday-life climate activists.

The everyday-life climate activists

Sustainable Lives is built upon the idea that local activism can be a response to global challenges. The slogan of the organisation is 'reduce the ecological footprint and raise the quality of life', which highlights the focus on sustainable activities and climate change-mitigating solutions as well as the importance of everyday life in local communities (Bærekraftige liv 2020b).[3] In this way, the connection between time and place is highlighted symbolically in the slogan of the organisation, time represented by the future which is presumed ruined by too deep an ecological footprint, and place by the quality of everyday life in a local neighbourhood.

The first Sustainable Lives group was established as a network organisation around 2008 in Landås, a suburb of Bergen, Norway.[4] In the past twelve years, the 'Sustainable Lives, Landås' group has had organisational and political success. The organisation has been appointed a seat in the National Climate Council of Norway; it has fundraised for two part-time employees, and it has established 'Sustainable Lives, Norway', which functions as an umbrella organisation to all of the existing groups. At present, 23 local groups throughout Norway are registered as Sustainable Lives groups, the majority of which (fifteen) are based in and around the greater Bergen area (Bærekraftige liv 2020a).[5] The activities of Sustainable Lives consist primarily of workshops in areas such as upcycling, vegetable gardening, or bicycle-repairing as well as social activities, including the recycling of clothes, gardening, and restoring the old house which houses the organisation, and, in Landås, organising an annual neighbourhood festival focussed on sustainability. The organisation roots itself in the everyday life of primarily middle-class suburban neighbourhoods. Norway figures on the Forbes top ten list of the richest countries in the world and Bergen is one of its largest cities, which of course sets the scene for a specific kind of privileged life. The neighbourhood consists mostly of single-family detached houses and even though there are some apartment buildings in some parts of this area, most of the activists do not live in those. As a network organisation, Sustainable Lives inscribes itself in the everyday-life activist landscape of activities which follow a general austerity trend towards making everyday life more sustainable, i.e. urban gardening or farmers' markets (Istenic 2018, 35). The organisation should be seen as part of a global organisational trend and is in many ways very similar to the UK-based 'Transition' movement (Transition Network 2019b).[6] Even though there are only vague official connections between 'Transition' and Sustainable Lives (Transition Network 2019a),[7] there are many parallels in the rhetorical strategies of the organisations regarding the future threat of climate change. The prime motivation of both organisations is to rearrange our lives in accordance with the assumption that climate change is going to change everyday life all over the world in fundamental ways.

It could be seen as a paradox that to ensure that our own existence continues in a somewhat similar way to the present, we must change our way of life. This reflects the inherent tension between stability and change which is present in this type of activism. Neither organisation demands radical civilisational or political changes, such as a life altogether without electricity or the removal of neoliberal capitalism, even though 'Transition' includes more radical initiatives than Sustainable Lives, such as the printing of local currencies (Noel 2012). The explicit and articulated aim of both organisations is to change the everyday lives of everyday people (in the Western world), and the expectation is that such a change will affect life on a grander scale in the climate-changed future in some way or another as well. In this way, the importance of everyday-life practices are emphasised, and it is evident that the ambition of activists in these organisations is to influence global changes with local practices rather than change global practices through global political decisions. This can be seen as a sort of bottom-up approach to the desired societal effects of climate change mitigation.

An additional analytical link between the Norwegian and the international organisation is that the terms 'sustainability' and 'transition' both can be seen as temporal and dynamic denominators pointing towards the future. The meaning of the term transition in the 'Transition' movement reflects a wish to prepare local communities for the climate-changed and oil-independent future, and to build resilient local communities that are equipped to deal with the transition into the societies of the future. The sustainability of Sustainable Lives entails the planning of a viable everyday life and a reduction of the climate impact of such a life. In both organisations, the presumption is that human life in the present can be organised in a way that prepares it for existence in a climate-changed future.

'Gjenkunning' – to regain knowledge and abilities

To Sustainable Lives, the concept of gjenkunning plays an important role in imagining what this sustainable climate-changed future contains. The word gjenkunning is not really a word in itself – at least not in Norwegian. It doesn't exist in Norwegian dictionaries, probably because it is a tentative translation of the English term 're-skilling'. The first part of the word – *gjen-* – is a translation of the English 're-' and refers to a repetition of the latter part. This aspect of repetition is not without significance as it adds a temporal and cyclical aspect to the concept. *Kunning* is more difficult to define as it is not simply a straightforward translation of 'skilling'. 'Skill' is the ability to use one's knowledge effectively and readily in execution or performance (Merriam-Webster 2019), and 'skilling' can be used to refer to training workers for a particular task (Lexico 2019). By using knowledge, execution, performance, and training, 'skilling' becomes a denomination of activity, of doing. Kunning is a fusion of the noun: a *kunning*, which describes a person who is very knowledgeable or skilled (Norsk Ordbok 2019), and the verb *å kunne*, which means to know and to be able to, but has the additional meaning 'to have' or 'to be given the opportunity to' (Språkrådet 2020). The 'kunning' of gjenkunning can be interpreted to mean something along the lines of 'to know, to know about, plus to be able to, and to have the opportunity to do something'. Gjenkunning, then, is a word that means to regain knowledge or ability and to have the opportunity to use this knowledge or ability. This is not as directly activity-oriented as 're-skilling', even though it indicates an expectation of the opportunity to put a regained knowledge to use – to both know and be able in practice.

Gjenkunning is not a very widely used word in Norwegian. In a Google search, it appears only ten times, which can be seen as being on the brink of existence in this day and age. Furthermore, all hits have a connection to Sustainable Lives, most of them as the hashtag *#gjenkunning* used on social media platforms (Instagram and Facebook). Nevertheless, gjenkunning can be seen as a semantically dense window onto the worldview behind the activities of Sustainable Lives. Even though the word is barely used and was actively constructed, it is explicitly used and explained in the context of Sustainable Lives and the connotations of it permeates the organisation, as shall be discussed below.

An explanation of gjenkunning figures in an article written by one of the founders of Sustainable Lives in a magazine called *Harvest*, a publication that focusses on 'nature, humans, and environment'.[8] This, rather long, quote serves to introduce the essential aspects of the concept of gjenkunning as presented by the organisation itself:

> Why do we grow plants in our gardens that we have to warn children against eating? Why can't they eat everything? What happens to us when the knowledge of whether the weeds right outside our door are edible or not disappears? Doesn't it make us less rather than more free, despite what former generations were taught? How do we activate the wealth of knowledge that a neighbourhood holds collectively? Someone knows that the linden hedge is edible, others which roots from the forest are exquisite in a wok. Someone has years of experience with vegetable gardens in exactly your neighbourhood and knows precisely which plants will flourish in the soil and climate in your street. Others hunt, chop meat, gather wild herbs, preserve, and pickle like naturals. These are valuable forms of local knowledge and resources that must be discovered and invited into the open. In English this is 're-skilling'. We call it 'gjenkunning', everything that people knew before that we must learn anew.
>
> (Tvinnereim 2019)[9]

In this quote, the gardens of the present 'we' are described as dangerous because of the lack of knowledge of which plants to eat, but also fundamentally absurd in the puzzlement of the fact that anyone would grow plants that are poisonous. Luckily the knowledge of the edibility of the linden hedge, of how to forage for food in the forest, to hunt, or to make preserves, does exist among the residents of the suburban neighbourhood in Bergen. The knowledge is accumulated through former generations and based on years of experience but lacks the practical aspect of transmission to present-day neighbours. Sustainable Lives wants to bring this knowledge into the open. By stressing the fact that it is a neighbour, maybe from another generation, who possesses the knowledge of which leaves are edible or which plant will flourish in exactly this soil, the '-kunning' of 'gjenkunning' is simultaneously placed in the past by referring to 'years of experience' and in the present by describing it as something which must be learned anew; the past and the present is thereby connected in the accumulated knowledge that must be discovered. Concurrently, the knowledge is located in a very specific place: this specific neighbourhood, Landås in Bergen. The latter part makes the perception of this knowledge very local and situated. Furthermore, the local neighbourhood is placed in a larger global context by referring to the climate of the neighbourhood, thus placing Landås firmly on the ground, but also into a larger context of climate discourse. In this way, gjenkunning becomes a point of intersection for the spatial and temporal scales of local/global and past/present/future.

Finally, I find that there is a cultural expectation of stability and a connection to a stable past in this presentation of gjenkunning. The accumulated pool of local knowledge, of 'everything that people knew before', is simply waiting to be 'discovered and invited into the open'. And the revitalisation of this pool of knowledge, agriculture, and handicraft is, by being situated in the context of Sustainable Lives, presented as part of local everyday-life climate activism.

It is no coincidence that the Norwegian organisation uses a designation such as gjenkunning. The word is an example of the inspiration that the organisation gets from the 'Transition' movement mentioned above. 'Transition' has launched the concept of 're-skilling' into the sphere of climate activism by introducing 'The Great Re-Skilling', which is step eight of 'The Twelve Steps of Transition' (Hopkins 2008, 148):

> I believe that one of the main factors contributing to the sense of panic [...] is the realization that we no longer have many of the basic skills our grandparents took for granted. One of the most useful things a Transition Initiative can do is to offer widely available training in a range of these skills.
>
> (Hopkins 2008, 166)

In this Transition version of 're-skilling', there is an emphasis on the loss of skills and the need for a renewed training in and transmission of these skills. The skills are established as 'basic' and taken for granted and the quote introduces previous generations as a resource with the nomination of 'grandparents' as bearers of the necessary skills. The resonance of this great 're-skilling' can be found in several layers of the gjenkunning of Sustainable Lives.

The lack of transmission of traditional skills and knowledge

In line with the slogan mentioned above ('We aim to reduce the ecological footprint and raise the quality of live'), a majority of the activities in Sustainable Lives are focussed on the production of and increased use of local commodities to reduce the CO_2 costs of transportation. Workshops on 'How to grow vegetables in your garden' (Bærekraftige liv Landås 2020)[10] or 'Wild herbs and how to use them' (Matskogen på Landås and Bærekraftige liv Landås 2020)[11] are examples of activities of this kind.

The framing of increasing the production of local commodities is part of the context of the concept of gjenkunning. The idea that local communities have to be self-sufficient in some form or another to survive in the climate-changed future plays a pivotal role in the activities of Sustainable Lives. The notion is that self-sufficiency will secure the resilience of local communities in an unknown, but surely different, future, albeit not in a way that presupposes an all-encompassing and destructive catastrophe. In Sustainable Lives, a presumption that the knowledge of how to be resilient existed in the past is added explicitly. Embedded in the term 'gjen-' is a cyclic longing for a return to past modes of life and the 'everything that people knew before' in Sustainable Lives is closely connected to an imagined self-sufficient past and the transmission of knowledge of this past. In an interview, one informant answered the question of why she has chosen to be involved in the activities of Sustainable Lives by saying, ‘My children do not know how to gut a fish’, going on to say that they do not know because she hasn’t taught them since she doesn’t know how to gut a fish herself. The informant’s reasons to be a part of an organisation that focusses on gjenkunning is both the lack of skill in future generations (her children) and her own inability to transmit the knowledge of the skills because she doesn’t have them either. It should be noted that Bergen is a coastal city and that fish-gutting is perceived, by the informant, as a fairly normal skill to transfer to your children.

The lack of transmission of skills plays a prominent role in the framing of gjenkunning. The quote below is from an article in a magazine entitled *Bærekraftige Liv* (Sustainable Lives). The magazine represents the organisation, containing articles that feature and are written by volunteers and employees. The headline is simply ‘Gjenkunning’:

> Based on my genes I should be anything but useless, since most of us, including me, are the descendants of an abundance of herring-eating entrepreneurs with enough vitality to survive by hanging on to the naked cliffs by the sea – and have a fire burning at the same time. Why can’t I do what they did? The answer is simple: nobody taught me how.
>
> (Koehler 2016, 34)[12]

With words such as *genes* and *descend*, and an image of vigorous Norwegians clinging to the naked cliffs, the sense of belonging is described in almost biological terms, but can here be seen as a way of connecting to a notion of generation. The sum of this generational continuity of knowledge was lost just a generation or two ago, when the prior generation to the writer’s own did not transfer their knowledge to her generation, thus making her unable to transfer it to her children. The quote describes all former generations of Norwegians as basically the same people connected by genes and nature and all present Norwegians – ‘most of us’ – as descendants and thus connected directly to both the knowledge and the abilities mentioned above. In this understanding, the lack of transmission of knowledge in the past makes gjenkunning a mode of reconnecting to the perceived thousands of years of skill and knowledge developed by the Norwegian predecessors. The quote above continues:

> For centuries, people have gained immense amounts of knowledge on how nature and natural materials work, and they have made sure to transmit this knowledge to their children and grandchildren to ensure them as good a life as possible. But after the industrial revolution in the beginning of the nineteenth century, more and more of this knowledge has been lost.
>
> (Koehler 2016, 35)[13]

This quote primarily establishes an understanding of a coherent past which ranges from centuries of connectedness to nature through the industrial revolution to the previous generation. This makes the past an indeterminable, unchangeable 'before'. The connection to this before is simultaneously destroyed by the industrial revolution and by generations immediately prior to the generational point of view of the text. This highlights a view of the premodern past as unchangeable and it establishes a connection to the stable past through generation and family. The cyclic understanding of temporality which lies embedded in the concepts of generation and family can, in the words of Tamara Hareven, be called family time. Family time has the ability to transcend the life experiences of the individual and combine time in a different way than a linear chronology does. Contrary to the elements of development and change in linear time, family time relies on repetition and stability, exemplified by the expectation of knowledge transmission to children and grandchildren. Hareven uses family time as a description of a connection point between history and the experiences of the individual (Hareven 1977, 59–61). In Sustainable Lives, ancestors and generation become the context of the continuity of time and gjenkunning emphasises the need to uphold this continuity.

Gjenkunning as traditionalisation

The use of the term gjenkunning can be seen to express a cultural expectation of stability via a connection through generations to an imagined stable past. Thus, the concept of tradition offers an important way in which to analyse the practice of gjenkunning as a connection to the past. An expectation of the probable future based on a cyclic sense of time is re-established by traditionalising certain contemporary practices. As mentioned, the process of ascribing the characteristics of tradition – continuity, stability, inheritance – to contemporary practices can be described as traditionalisation. In this process, tradition as a quality or a value is ascribed to certain practices, ideas, or art forms, and the traditional is seen as a value in itself (Eriksen and Stensvold 2002, 74, 86–87). Values are in this instance not meant to describe either positive or negative characteristics; tradition is used as a value in and of itself. Traditions are furthermore used to state the continuity between past, present, and future (e.g. Glassie 1995). Traditions and the use of tradition as a value make it possible to imagine a somewhat recognisable society in the future because traditions per se point to a continuity in time, in line with the introductory assumption that traditions are the creation of the future out of the past, as stated above with reference to Henry Glassie (Glassie 1995, 409).

In using gjenkunning as a designation connected to climate change mitigation, Sustainable Lives subscribes to an understanding of traditions as interlinked chains of values and practices that are transmitted by the bearers of these traditions who undertake the responsibility of the transmission. Folklorist Dorothy Noyes describes tradition as a task to be done, and she underlines that it is the responsibility of the holders of tradition to find someone willing and competent to do the job (Noyes 2017, 110). Sustainable Lives can be seen as positioning themselves, as part of the discourse of the organisation, to be the conveyors of tradition and the menders of the broken chain, actively seeking out the holders of tradition and inviting them into the community. The unbroken, or mended, chain of tradition bearers is perceived as the guarantee that the traditions will arrive unchanged from the past to the present and will be transmitted to the future in the same fashion. I find this everyday understanding of tradition in various descriptions of gjenkunning. The quote below, taken from the article 'Gjenkunning' mentioned above, begins with the writer stating her own inadequacy in traditional skills:

> Did you – as I did – scratch your head and wonder how to put up a fence without it collapsing again during the night.... Maybe you really wish that you had remembered how your grandfather practiced net fishing or cleaned the fish, because you wanted to take your own children out in a boat in the weekends [...].
>
> (Koehler 2016, 34)[14]

In this quote, the writer describes her own lack of skills and places her own generation as the 'missing link' in the transmission of knowledge. She wishes that she paid attention to how the previous generation practised the knowledge because this would have enabled her to transmit it to her children. She describes this as a vital deficiency in her everyday life. The activities described (putting up a fence, fishing, and sailing) are all traditionalised by being placed in the past and being linked to previous generations through the grandfather who is portrayed as a carrier of this vital knowledge.

Gjenkunning is furthermore stressed as a meaningful thing to do in everyday life in the following quote from the gjenkunning article, which is about the growing of potatoes and the preparation of meat. The text is placed beneath three photos: one shows a pair of hands holding some potatoes that have just been dug up from the field that shows in the background. Dirt clings to both hands and potatoes. It is a colour photo and both the green of the field and the brown of the dirt stands out next to the other photos on the same page, which are both in black and white (Figure 3.1). One is of a small child looking at a simulated cow's udder (it looks like a pair of plastic gloves on a rack) with drops of fluid lingering at the tips. The child has one finger in the mouth and looks bewildered and a little cautious. The last black-and-white photo shows a middle-aged woman carving some sort of meat with a large knife. The text reads:

> **Grow your own potatoes.** It's not really all that difficult to do and they taste wonderful. In the past it was normal to know how to **quarter an animal**. It makes sense to use the whole animal.
>
> (Koehler 2016, 35)[15]

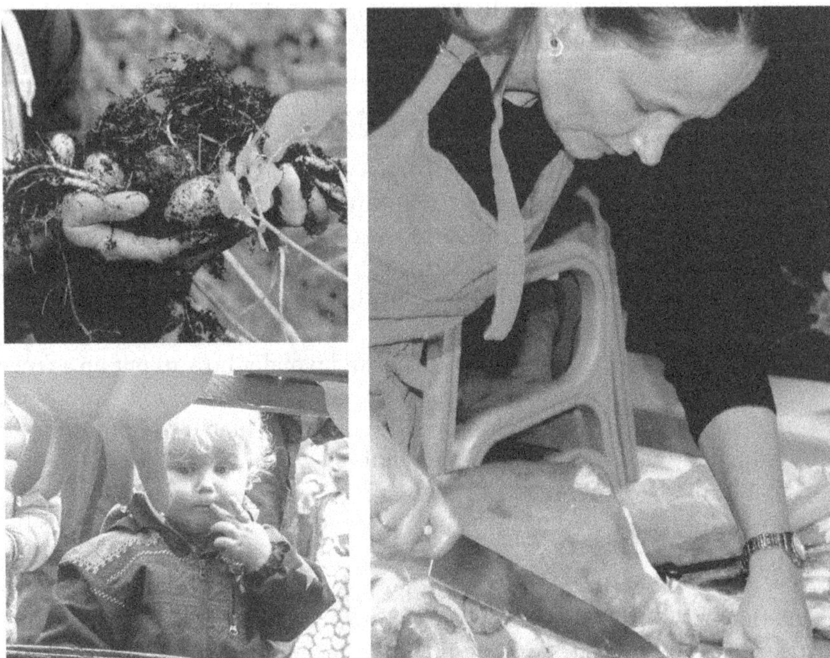

Dyrke egne poteter. Ikke så veldig vanskelig å gjøre, og smaker herlig. Partering av slakt var vanlig kunne før. En meningsylt ting å gjøre for å utnytte seg av hele dyret. Hm...mon tro hva denne gutten på LandåsFest lærte om kuer, melk og kjærlighet.

Figure 3.1 Facsimile from Sustainable Lives Magazine.

It is important to note initially that potatoes are a common vegetable to grow nowadays in most parts of Norway if you have a vegetable garden and that potatoes are part of the ordinary Norwegian diet. The quartering of meat, on the other hand, is not something you would do at home and not something most people would know how to do. By presenting the growing of potatoes and the quartering of meat as part of the same gjenkunning, and as 'normal in the past', the writer, first, states a very urban point of view removed from any kind of farming and, second, creates a certain image of the old days linked to husbandry and self-sufficient smallholding. The text constructs the growing of potatoes and the quartering of meat as valuable practices which are placed in an ideal society, both in the past and in the future. The activity of growing potatoes is traditionalised by presenting it as a positive and rewarding activity, and by the stressing that 'it's not really all that difficult to do', which underlines the perception that it is not a common activity in the present and that one needs to relearn how to do it. The knowledge of growing potatoes and quartering meat is tied to a vaguely defined nostalgic past with the assumption that these used to be common skills. It is a given that this was normal 'in the past' in the same way that it is assumed that people will think that it is difficult in the present, hence the need to stipulate that it is not.

The writer further encourages the beginning of gjenkunning before it is too late, and traditional knowledge forms will have disappeared:

> It seems like a good idea to start before we have to – and not least: before the knowledge is lost. A lot of people have realised this already, so many, in fact, that we can say that the Great Gjenkunning has started.
>
> (Koehler 2016, 35)[16]

By using the designator 'the Great Gjenkunning', the writer connects Sustainable Lives directly to the international 'Transition' movement by mirroring Hopkins's eighth step of transition: 'the Great Re-Skilling'. 'The Great Gjenkunning' is even capitalised, which is not grammatically correct in Norwegian, underlining both the interorganisational connection and the conceptualisation of gjenkunning. Furthermore, by using capitals and writing it in definite singular form, the author marks out the Great Gjenkunning as an all-encompassing transformation. In that regard, it could be regarded as a collective singular (cf. Koselleck 2004). Like 'The Revolution' or 'The Reformation', it is more than just an event, and appears as a historical force in its own right and with its own intrinsic rationale (cf. Koselleck 2004, 50). Koselleck has demonstrated that re-volution and re-formation entail a double temporality, which can be found in gjenkunning as well. The two concepts in Koselleck's analysis originally indicated a return or a restoration of an original order, as indicated by the prefix 're-'. As collective singulars, their meanings changed. Both concepts came to imply the creation of something entirely new and were even ascribed a nearly autonomous historical agency. A similar duality is also embedded in 'the Great Gjenkunning' as it explicitly refers to past practices, yet at the same time points towards a new way of living.

A nostalgic longing for tomorrow?

On the everyday, general level, nostalgia is seen as an emotionally laden longing for the past and as a filter which gives the past a rosy colour and edits out or reinterprets its negative aspects (Tannock 1995, 457; Boym 2001). This kind of nostalgia has been labelled simple nostalgia (Boym 2001, xviii, 41–42; Wilson 2015, 480). To label something as nostalgic at this level implies an air of incredibility and naivety to its image of the past. It's only nostalgia (Wilson 2015, 480). But nostalgia can also be viewed as an important component of the way past, present, and future are connected through tradition and generation in Sustainable Lives. On an analytical level, nostalgia can be seen as a concept which 'encapsulates a recalling of the past, in the present, with the potential of anticipating the future', as suggested by social psychologist Janelle Wilson (Wilson 2015, 478). And as a cultural phenomenon, nostalgia describes, in the words of Swedish cultural historian Karin Johannisson, a longing for a space of existence (Johannisson 2001, 34). In this way, Sustainable Lives is clearly a nostalgic organisation at its core, with gjenkunning working to recall the past in anticipation of a space of existence in the climate-changed future. But establishing how the nostalgia of Sustainable Lives works is not a simple task – neither in a theoretical nor in an empirical way.

The texts that I have quoted above show a view of a past that is multifaceted. It is a past of husbandry in smallholdings in rural Norway, a life in close connection to wild vegetation with knowledge about food production and nature accumulated through thousands of years. It is a past, maybe portrayed in somewhat broad strokes, in which people knew how to grow potatoes, what vegetables would flourish locally, and what wild plants were to be eaten or not to be eaten. This 'thousands of years of knowledge', based on the ability to 'hang on to the naked cliffs', was passed on to the next generations as a part of everyday life. Is this simply a nostalgic view of the past? Yes, maybe it is, because it does idealise the past, but it also takes a detour into a more survival-focussed kind of nostalgia. Progressive nostalgia is a term introduced by heritage scholars Laurajane Smith and Gary Campbell to capture the selective choosing of elements from the past as a somewhat conscious process with the explicit purpose of highlighting these elements and how they are influencing political and social agendas in the present that point to the future (Smith and Campbell 2017, 612–613). Smith and Campbell describe progressive nostalgia as a way of connecting qualities that are valued in the present to the past. Qualities such as hard work and discipline are seen as achievements and are remembered in active and unashamedly emotional ways – ways that you could call nostalgic. This emotion is used as a way of explicitly directing the nostalgia towards the future by setting a politically progressive agenda for future living and working conditions based on the assumption that the past was not perfect, but that it is the precondition of the present (Smith and Campbell 2017, 613).

The nostalgia of Sustainable Lives does not include the explicit depiction of the past as hard and difficult, a similar exclusion to the instances that Smith and Campbell describe, which are primarily connected to industrial heritage and associated progressive nostalgic values such as hard work and community camaraderie (Smith and Campbell 2017, 615). But when the writer of the gjenkunning article describes her ancestors as vigorous 'herring-eating entrepreneurs', hanging on to the naked cliffs and simultaneously keeping a fire burning, she points not to an ideal golden age but to a past in which her ancestors had to be tough to survive. The article does not merely offer a reinterpretation of negative aspects to make them appear positive but instead places a progressive-nostalgic emphasis on the necessary skills to survive under harsh conditions. This points to an image of a climate-changed future which might bring harsh conditions, but for which gjenkunning will help climate activists be prepared.

The gjenkunning of Sustainable Lives

Gjenkunning is a vital part of the climate change mitigation of everyday climate activists in the Sustainable Lives organisation in Bergen, Norway. The imagining of the past, as well as the future, is closely connected to the practices of Sustainable Lives. Through a traditionalisation of certain practices, tradition becomes an argument used to emphasise continuity. In the contradictions of stability and change encompassed by the concept of tradition and the processes of traditionalisation lies a temporal complexity which simultaneously connects Sustainable Lives to the past and to the future. Despite a seemingly presentist focus, which blames an indeterminable past and fears the future, it becomes plausible to imagine an everyday life in a climate-changed future. The future is imagined with reference to a joint nostalgic past connected to the family time of generation and traditionalisation, and anchored in the practices of everyday-life climate activism. By constructing the term gjenkunning and attributing values like continuity and home to practices associated with the term, the connection between past, present, and future is established and it is made plausible that the future is in fact going to exist, even though its climate will be changed, and furthermore that it will exist in a form that resembles the past and the present. Gjenkunning, then, can be seen as a strategy to make the climate-changed future about continuity and not about disruption. It is a way for activists to gain control and manage the risks of the future. By analysing gjenkunning as a specific aspect of everyday-life climate change mitigation, I have shown that in this case the past is also intrinsically entangled with the future and the present, and that these entanglements are established through the imagining of a climate-changed future and what a possible, and maybe even good, community looks like in this future.

Notes

1 *Bærekraftige Liv* in Norwegian will be translated to Sustainable Lives (my translation).
2 This text corpus is a part of the empirical material in my PhD thesis, which also consists of qualitative interviews and observations, primarily in connection with the Landås group of Sustainable Lives.
3 All aktivitet har som hovedmål å bidra til 'redusert økologisk fotavtrykk og økt livskvalitet'. barekraftigeliv.no/inspirasjon/mening-handling-og-håp, accessed February 19, 2019.
4 Bergen has approximately 280,000 citizens and is the second largest city in Norway.
5 barekraftigeliv.no/arrangement/2018/det-er-i-nabolaget-verden-forandres/#map. The fifteen groups in the Bergen area are as follows: Hellen, Kalland/Stend, Kronstad, Landås, Løvstakken, Nattland/Sædalen, Nordnes, Os, Fyllingsdalen/Søreide, Skansen, Sandviken, Sydnedhalvøen, Åsane, Fana, and Minde. Additionally, there are four groups in and around greater Oslo: Holmlia, Mortensrud, Røyken, and Nesodden. The rest of the groups are in Bø in Telemark, Voss, Trondheim, and

Stavanger. The majority of the groups appear to be active, though a few only have messages of 'no further activities planned' posted, accessed January 20, 2014.

6 Transition has approximately 330 groups registered in the UK, 300 in the USA, 4 in African countries, and 15 in South American countries (plus a few in other parts of the world). transitionnetwork.org/transition-near-me/, accessed February 26, 2019.

7 'Sustainable Lives, Landås' does figure as a Transition group on the Transition Network webpage, but with no further content on the page, transitionnetwork.org/transition-near-me/initiatives/landas-transition-initiative-baerekraftige-liv-pa-landas, latest accessed April 3, 2019. Some of the key persons in the organisation have participated in educational activities in 'Transition' (personal information in interview).

8 www.harvestmagazne.no/om-harvest

9 All of the originally Norwegian quotes used as empirical material are my translation.

> Hvorfor har vi anlagt hager og fellesområder der vi må advare barna mot å spise enkelte planter? Hvorfor kan ikke alt spises? Hva skjer med oss når kunnskapen om hva som kan spises av «ugresset» rundt beina våre dør ut? Gjør ikke det oss mindre heller enn mer frie, slik de siste generasjonene er opplært til å tro? Hvordan henter vi ut, og sprer videre, den store rikdommen av kunnskap som et nabolag kollektivt besitter? Noen vet at lindehekken kan spises, andre hvilke røtter i skogen som er ypperlige i wok. Noen har mange tiårs erfaring med kjøkkenhage i akkurat ditt nabolag, og vet nøyaktig hvilke planter som klarer seg best i jordsmonnet og med de klimatiske forholdene i din gate. Andre jakter, parterer, sanker, safter og sylter med den største selvfølgelighet. Slik kunnskap er verdifull kompetanse som må oppdages og inviteres frem som en lokal ressurs. På engelsk heter det «re-skilling». Vi kaller det «gjenkunning», alt det folk kunne før som vi må lære oss igjen.

10 *Kjøkkenhagekurs*, onsdag d. 22. mai 2019, facebook.com/events/2249999901723812/, accessed January 20, 2014.

11 *Ville vekster og bruk av dem*, søndag d. 26. mai 2019, facebook.com/events/343403603034033/, accessed January 20, 2014.

12 'Etter genene å dømme burde jeg være alt annet enn ubrukelig, siden de aller fleste av oss, meg inkludert, nedstammer fra en anselig mengde sildespisende, selvbyggere med livskraft nok til å overleve ved å klore seg fast til nakne klipper ytterst ved havet – med ild i peisen. Hvorfor kan jeg ikke det de kunne? Svaret er enkelt: Ingen har lært meg det'.

13 'I årtusenenes løp har mennesker tilegnet seg enorm kunnskap og erfaring om hvordan naturen og naturmaterialer fungerer, og sørget for å overlevere summe av dette til barn og barnebarn, slik at de kunne leve så godt som mulig. Men etter den industrielle revolusjonen startet på 1800-tallet har mer og mer av denne kunnskapen gått tapt. Vi er nødt til å ta ansvar, og det er det mange som føler på'.

14 'Har du, som meg, også stått og klødd deg i hodet og lurt på hvordan i all verden du skal starte for at få satt opp et gjerde uten at det ramler ned bit for bit over natten? Eller vært full av beundring over en gammel speiderkar som vet nøyaktig hvilket magisk mønster vedkubbene må legges i for at de skal ta fyr når førsteklassingene er på grilltur i en fuktig skog? Kanskje du virkelig skulle ønske at du husket hvordan bestefar satt garn og sløyet fisk, fordi du gjerne ville tatt dine egne barn med ut i båt i helgene, for å gi videre noe av de samme gode stundene du opplevde da du var liten?'

15 '**Dyrke egne poteter**. Ikke så veldig vanskelig å gjøre og smaker herlig. **Partering av slakt** var vanlig kunne før. En menings[f]ylt ting å gjøre for å utnytte seg av hele dyret. [...]'. Bold in quote.

16 'Det virker som en god idé å starte før vi blir tvunget til det – ikke minst, før kunnskapene går tapt. Det er faktisk så mange som har innsett dette allerede, at vi kan si at den Store Gjenkunningen har startet'.

References

Bærekraftige liv. 2020a. 'Her finnes Bærekraftige liv'. https://www.barekraftigeliv.no/bli-med#map.

———. 2020b. 'Mening, Handling og Håp'. https://www.barekraftigeliv.no/inspirasjon/mening-handling-og-håp.

Bærekraftige liv Landås. 2020. 'Kjøkkenhagekurs'. Facebook. https://www.facebook.com/events/2249999901723812/.

Barr, Stewart, and Andrew Gilg. 2006. 'Sustainable Lifestyles: Framing Environmental Action in and Around the Home'. *Geoforum* 37: 906–920.

Bauman, Richard. 2004. *A World of Others' Words: Cross-Cultural Perspectives on Intertextuality*. Malden [etc.]: Blackwell.

Bauman, Zygmunt. 2017. *Retrotopia*. Cambridge: Polity Press.

Boym, Svetlana. 2001. *The Future of Nostalgia*. New York: Basic Books.

Cashman, Ray. 2006. 'Critical Nostalgia and Material Culture in Northern Ireland'. *The Journal of American Folklore* 119 (472): 137–160.

Dowling, Robyn. 2010. 'Geographies of Identity: Climate Change, Governmentality and Activism'. *Progress in Human Geography* 34 (4): 488–495.

Eriksen, Anne, and Anne Stensvold. 2002. *Maria-Kult og helgendyrkelse i moderne katolicisme*. Oslo: Pax forlag.

Glassie, Henry. 1995. 'Traditon'. *The Journal of American Folklore* 108 (430): 395–412.

Handler, Richard, and Jocelyn Linnekin. 1984. 'Tradition, Genuine or Spurious'. *Journal of American Folklore* 97 (385): 273–290.

Hareven, Tamara K. 1977. 'Family Time and Historical Time'. *Daedalus* 106 (2): 57–70.

Hopkins, Rob. 2008. *Transition Handbook : From Oil Dependency to Local Resilience*. Totnes: Green Books.

Istenic, Sasa Poljak. 2018. 'Green Resistance or Reproduction of Neoliberal Politics'. *Ethnologia Europaea: Journal of European Ethnology* 48 (1): 34–49.

Johannisson, Karin. 2001. *Nostalgia. En känslas historia*. Stockholm: Bonnier.

Koehler, Vibeke. 2016. 'Gjenkunning'. *Bærekraftige Liv*, 32–36.

Koselleck, Reinhart. 2004. *Futures Past: On the Semantics of Historical Time*. New York, Chichester, West Sussex: Columbia University Press.

Kverndokk, Kyrre. 2020. 'Talking about Your Generation: 'Our Children' as a Trope in Climate Change Discourse'. *Ethnologia Europaea: Journal of European Ethnology* 50 (1): 145–158

Lexico. 2019. 'Skill'. https://www.lexico.com/en/definition/skill.

Matskogen på Landås, and Bærekraftige liv Landås. 2020. 'Ville vekster og bruk av dem'. https://www.facebook.com/events/343403603034033/.

Merriam-Webster. 2019. 'Skill'. https://www.merriam-webster.com/dictionary/skill?

Noel, Longhurst. 2012. 'The Totnes Pound: A Grassroots Technological Niche'. In *Enterprising Communities: Grassroots Sustainability Innovations*, edited by Anna Davies, vol. 9, 163–188. *Advances in Ecopolitics*. Bingley: Emerald Group Publishing Limited.

Norsk Ordbok. 2019. 'Kunning'. http://no2014.uib.no/perl/ordbok/no2014.cgi?soek=kunning#ariadne=[[%7C149367%7C, 0,%7Ckunning%7C]].

Noyes, Dorothy. 2017. *Humble Theory: Folklore's Grasp on Social Life*. Bloomington: Indiana University Press.

O'Brien, Karen, Elin Selboe, and Bronwyn M. Hayward. 2018. 'Exploring Youth Activism on Climate Change: Dutiful, Disruptive and Dangerous Dissent'. *Ecology and Society* 23 (3): 42.

Smith, Laurajane, and Gary Campbell. 2017. 'Nostalgia for the Future: Memory, Nostalgia and the Politics of Class'. *International Journal of Heritage Studies* 23 (7): 612–627.

Språkrådet. 2020. 'Kunne'. Bokmålsordboka | Nynorskordboka. https://ordbok.uib.no/perl/ordbok.cgi?OPP=+kunne&ant_bokmaal=5&ant_nynorsk=5&begge=+&ordbok=begge.

Tannock, Stuart. 1995. 'Nostalgia Crtitique'. *Cultural Studies* 9 (3): 453–464.

Transition Network. 2019a. 'Landås Transition Initiative - 'Bærekraftige Liv På Landås''. https://transitioninitiative.org/initiatives/landas-transition-initiative-baerekraftige-liv-pa-landas/.

———. 2019b. 'Transition Near Me'. https://transitionnetwork.org/transition-near-me/.

Tvinnereim, Agnes. 2019. 'Hva om det er 'vi' som er 'noen'?' *Harvest Magazine*. https://www.harvestmagazine.no/pan/hva-om-det-er-vi-som-er-noen-1.

Wilson, Janelle Lynn. 2015. 'Here and Now, There and Then: Nostalgia as a Time and Space Phenomenon'. *Symbolic Interactionism* 38 (4): 478–492.

4 In the shadow of apocalyptic futures

Climate change as a cultural trope in vernacular discourse

Camilla Asplund Ingemark

Climate change as a cultural trope in vernacular culture

What is the status of climate change as a cultural trope in vernacular culture, and how does it function? These are the guiding concerns of this chapter, which focusses on responses to qualitative questionnaires on climate change and weather from several Nordic countries. In line with the general argument of this book, I propose that climate change as a concept is capable of organising multiple temporalities, while at the same time being organised by certain other concepts of a similar nature, such as catastrophe or crisis (cf. Kverndokk 2017). These temporalities form chronotopes (timespaces) that have been absorbed into the vernacular texts I study, and I examine through careful textual analyses how these temporalities play out as those texts coordinate and present them.

While I am suggesting that chronotopes do in fact have an agency in and of themselves – deriving from the multiple prior contexts in which they have been used and affecting the accounts given by respondents to the questionnaires in ways that these respondents might not have anticipated – they are also filtered through the narrative voices of respondents. This means that the respondents imbue the chronotopes with their own meanings in order to negotiate their own positions in relation to climate change.

Accordingly, the present chapter proceeds from a discussion of the pertinence of the concept of chronotopes to vernacular texts on climate change, to an empirical investigation of how respondents position themselves vis-à-vis climate change through their use of chronotopes. I then turn to the chronotopic concepts that seem to organise vernacular conceptions of climate change – such as catastrophe and crisis – and how they affect the texts, ending with a prospective contemplation of whether these organising concepts will eventually switch places, with climate change displacing catastrophe and crisis as the primary concept.

The material on which this chapter is based consists of 288 vernacular texts in all, collected through qualitative questionnaires in Finland, Norway, and Iceland (in chronological order). In practice, I have selected a few responses that are sufficiently long and detailed to afford an analysis of multiple temporalities while being short enough to be possible to quote and analyse in full. I have attempted to maintain an equal distribution of men and women, and a variety of ages, as well as to represent different points of view and emphases.

The organisation of chronotopes: narrative voice and positioning

To begin with, my focus will be on chronotopes on the level of motifs: these can be combined in different ways in each text, and are often quite heterogeneous, as we will see, although some appear to recur more often than others. Frequently, they derive from media reports and can crystallise whole multi-episodic narratives in a fleeting mention (for an example, see Ingemark 2019 on sinking islands). While such chronotopes of single motifs are multifarious and heterogeneous, they are made into a more or less coherent whole by being filtered through the *narrative voice* (Genette 1983, 212–213; Prince 2003, 104) of the respondents to the questionnaires, who select motifs from this largely pre-existing repertoire of chronotopes and mould them in accordance with their own perspectives and purposes. I am interested in how this selection of motifs is carried out and why, as well as in how the respondents actively use them in speaking of climate change. It is this emphasis on the act of speaking or narrating in the texts that warrants the use of the concept of narrative voice, which is more concerned with the question of 'who speaks' than with those of 'who sees' (Prince 2003, 104).

As for the purposes to which these chronotopes are put, they can certainly vary. Some employ these motifs to argue for the existence of climate change, while others use them to deny it. When we look closely at the texts, we can often see that respondents are negotiating their positions vis-à-vis climate change as a topic, on the one hand, while simultaneously entering into dialogue with the authors of the questions, on the other hand. This double orientation leaves evident marks on the texts, and can be further complicated by a third orientation towards an imagined or anticipated (sympathetic) addressee in the case of climate change deniers who, perhaps for good reason, assume that the compilers of the questions do not agree with their own points of view. I adapt psychologist Michael Bamberg's notion of *positioning* (Bamberg 1997, 336–337) to explore how the respondents are negotiating their positions in the texts in these three ways.

Bamberg distinguishes between three levels of positioning, which are somewhat differently articulated from mine: (1) how characters in a narrative are positioned in relation to one another, (2) how narrators position themselves in relation to an audience, and (3) how narrators position themselves in relation to themselves (Bamberg 1997, 337). In this paper, I have replaced how characters in a narrative are positioned in relation to one another with how narrators position themselves in relation to the topic of climate change.

This is an adaptation to the special nature of discourse on climate change, which seems to require actively taking a stance for or against the existence of climate change; this implies a moral imperative to position oneself. As my primary focus is on how narrators position themselves in relation to an audience, the notion of positioning is complemented with literary critic Mikhail Bakhtin's concept of *addressivity* (Bakhtin 1986a, 280; 1986b, 69), i.e. the orientation of an utterance towards the anticipated reply of an imagined recipient (*addressee*). Bakhtin suggests that each utterance we make is already directed towards a response, and that the utterance is being shaped by this anticipated response. Thus, I use the concept of addressee to examine how respondents are orienting themselves towards this imagined recipient. Since it is difficult to disentangle positioning in relation to an audience from positioning oneself in relation to oneself, this aspect will also be discussed to some extent.

As the argument progresses, the focus will gradually move towards the generic and conceptual level of chronotopes. The top-most one is what we might call the chronotope of the questionnaire, which operates as a secondary genre absorbing other, simple genres into itself (Bakhtin 1986b, 61–62). The chronotope of this genre is rather difficult to define, and in this context, it might not be the most productive aspect to investigate either; the constituent genres are far more interesting to study. Suffice it to say that the folkloristic questionnaire in its current form can be defined as a genre through its particular use of narrative voice. Since the 1970s, qualitative questionnaires within the fields of folkloristics and ethnology have been designed to elicit personal narratives and reflections on everyday life, folklore and folklife, rather than impersonal, collective accounts of the folk traditions of a given geographical area as they were once wont to do (see e.g. Lilja 2016). The respondent is no longer expected to be a ventriloquist for a whole tradition but is supposed to speak in her or his own voice. When it comes to climate change, this expectation of personal accounts poses some challenges for the respondents compared to most other topics, or even to other parts of the same questionnaire, as in the Finnish case. It is easier to tell a story of a violent storm you experienced (see Marander-Eklund 2016), for example, than to turn your opinions on climate change into a story. That is probably why the responses I examine move more towards the 'personal reflection' end of the spectrum.

Further down the ladder of chronotopes, we find the conceptual chronotopes of catastrophe and crisis which have their own peculiar temporalities, and which can be easily integrated into many different genres (more on this below). Another prominent chronotope is that of deep time, which can be described as a *chronotopic model* of how the world is constituted. Asif Agha introduced this term using an example related to mine: namely, the struggle between Darwinists and Creationists, which is framed as a dispute between the competing chronotopes' 'evolutionary history' and 'biblical time' (Agha

2007, 322). This dispute is, surprisingly, largely absent in my material; nearly everyone seems to take deep time and evolutionary history for granted. Where they differ is in the perceived nature of this evolutionary history, and the scenarios for its future evolution. Those who are convinced of the existence of climate change and climate change deniers can be said to align themselves with different variants of the same chronotopic model – deep time – through a distinctly social process that also has social consequences. Agha writes:

> More generally, whether or not a chronotopic model is widely known, is felt to be legitimate, is uniformly accepted by those acquainted with it, or whether it fractionates into positionally entrenched variants, the process as a whole proceeds *as a social process* through modes and moments of participatory access to the model itself (i.e., through semiotic activities that unfold within participation frameworks) and through forms of alignment to *that* model (or variant) to which participants orient in some modality of response (registering uptake, maintaining its presuppositions, countering its features, proposing alternatives, etc.) through their own semiotic activities (Agha 2007, 322).

Climate change is the point at which the general consensus on the existence of deep time 'fractionates into positionally entrenched variants', and this social process relies on the fact that we all have participatory access to the chronotopic model of deep time through various means. In aligning with these fractionated variants, people respond to different aspects of the chronotopic model, favouring some features and suppressing others, and it is through 'their own semiotic activities', in this case their responses to questionnaires, that we have access to their thoughts.

A catastrophic future

As I first approached the task of examining temporalities in the questionnaires, I was despairing because I could see very little relating to temporalities in them. After a closer look, however, I realised that there was actually a plethora of temporalities; they were just not the ones I had expected to find. If climate change temporality is constituted through the entanglement of deep geological time and historical time as Dipesh Chakrabarty (2009) has argued, the general absence of explicit discussions of deep time in the responses initially seemed bewildering. Thus, deep time is usually discussed by proxy, and what these proxies are will be the focus of this subchapter.

In this response by a female respondent, born in 1970, to the Norwegian questionnaire, many different times are drawn upon to represent climate change, which functions as the organising concept (I have added the questions included in the questionnaire in square brackets to enable the reader to judge what she is responding to):

> [*What do you associate with man-made climate change?*] Decreasing biological diversity, constant news of species being threatened with extinction and of invasive species displacing local ones. Changes in the weather, greater instability, displaced seasons.
>
> [*Have you personally experienced some impacts of climate change?*] Well, can we know we are having [such consequences]? I think so, for instance I think that the very hot and dry spring/early summer we have had this year is due to a changed climate. And the unstable winters of the last twenty years. The many violent storms you read about. We have been protected from the most extreme thus far. Climate change influences my life partly because it has changed the way in which I live (changed travel habits and such things, cf. below) and partly because it worries me greatly. I have no optimistic or positive belief in the future, and that is chiefly due to climate change and humanity's attitudes to it.
>
> [*Do you do anything to reduce climate change?*] I travel less by air – for example, I prefer taking the train when it is manageable in terms of time and I abstain from travelling as far and as often as I wish. As for weekend trips to big cities, I'd rather go to cities possible to reach by train than those one must travel to by air, and I do not travel on more than 1–2 weekend trips per year. And if I need to travel for business to a place I find interesting, I try to stay a few days extra at my own expense to make one trip fill many functions.
>
> [*Where do you get information on climate change?*] I trust knowledge based on research and organisations working for change, whether they are state-run or not. I read the newsletters and websites of these organisations as well as daily papers and magazines etc.
>
> [*Do you believe there are other societal threats greater than climate change?*] No, but I think many of the consequences of climate change – climate refugees, migration, water and food shortage, loss of biodiversity/collapse of ecosystems, struggle for resources – will become very great threats to us, create very big problems.
>
> [*How do you envision the future?*] I'm a pessimist. I don't think we can manage to make the changes that are required. Humanity doesn't seem capable of organising itself well enough, arrive at a common understanding of how serious this is, so we cannot manage to change everything we do wrong today, and therefore it won't work; we can't make it turn around. I don't think we realise the gravity [of the situation], especially here in Norway where we are still so protected from the impact of climate change.
>
> Another thing is what should we substitute for today's global capitalist economy, since it destroys the Earth? How can we transition from a society such as the present one where we live off producing things and

> services for each other in a manner that uses up resources and destroys the environment, and where the goal is for us to consume more and more, and into something new that doesn't do that and still succeed in getting everyone a job and income? I can't understand it.
>
> So I believe we are moving towards a catastrophic future in which human society collapses and humanity is gradually ruined. How will we manage to get enough food for all of us, for example, when biodiversity is so reduced/destroyed that ecosystems break down? I don't think we can survive in a ruined world, and I hope we can't. I think it's terrible to see how egotistical our species is; we speak of plant and animal species that are invasive and destroy their new environment, but we have done it ourselves. We swarm the Earth and use up resources without any thought of tomorrow. Many times I say to myself that it was wrong of me to have children of my own, for what kind of world am I bequeathing to them? Their life will be worse than mine has been. For their sake, I hope they choose not to have children. Living with children in the world I see coming in the future seems impossible.
>
> (NEG 0263/00010. Woman, b. 1970)

As a response to the question of what she associates with the concept of climate change, she begins by appealing to a biological timescale of a – implicitly accelerating – loss of biodiversity, but quickly switches to the equally accelerating pace of global news coverage, which gives us ever new reports of species threatened with extinction or invasive species displacing local ones. Then she mentions fundamental changes in weather patterns, the increasing instability of the weather, and the temporal disruption of the seasons.

In the first question, respondents were invited to associate freely around the concept of climate change, and the replies frequently assume the form of lists, as in this case. These lists tend to condense the chief meanings of climate change for the respondent, often in at least partially experiential terms, as here, where changes in weather and in the seasons are a prominent theme. This topic recurs in the response to the second question pertaining to her personal experiences of the consequences of climate change. The Norwegian responses were submitted after the heat wave and concomitant drought in the Nordic countries in the summer of 2018 (cf. Chapter 5), and the respondent refers to it as one principal experiential meaning for her of climate change. However, she begins her answer by posing a question: 'Well, can we know whether we are having [such consequences]?' When she immediately replies in the affirmative, this makes it a purely rhetorical question, which serves to position her as a climate conscious person. It could also be interpreted as a strategy for creating a rapport with the authors of the questionnaire and establishing a sense of mutual understanding and agreement.

The respondent makes a clear distinction between her own experiences in Norway and the more distant problem of violent storms in other parts of the globe, events that she knows about through the media. The timescales of these phenomena are quite different; the heat wave and unstable winters are parts of a continuous flow of experiences in everyday life, albeit particularly highlighted and memorable ones, whereas the violent storms occurring elsewhere are transmitted to her only intermittently, in the fast pace and temporalities of news coverage.

It is at this point that the temporalities of travel first emerge as a theme, as a response to the question of whether climate change has had an impact on her personally. The topic recurs in her answer to the next question, whether she does anything to reduce such impacts. In between, we find the apocalyptic timeframe of an envisioned climate catastrophe, which prevents her from being confident in the possibility of a good future for mankind. The timescales of different types of travel are discussed, and the slower pace of train travel is preferred when it is 'manageable' in temporal terms. Acquiring new travel habits remains the main topic throughout the response to this question, and the respondent describes her attempts to limit travel in general and air travel more specifically. As before, the respondent seems to orient herself to the compilers of the questionnaire, seeking agreement with the appropriateness of her chosen course of action.

Altered travel patterns do indeed figure very prominently in the Norwegian material; several respondents also mention buying an electric vehicle recently, which reflects Norway's position as number one globally in electric cars sales per capita – in 2018, 31.2 percent of all new cars were pure plug-ins (OFV 2019), and this figure has risen since. Here it is interesting to note that none among the Norwegian respondents ever venture to question the legitimacy of continued oil production in Norway. This might be partially due to the way in which the queries are framed, since they are slanted more towards the respondents' personal agency than towards their opinions about the responsibilities of states and other public bodies. Yet even when such responsibilities are brought up, this aspect is passed by with silence.

Turning to visions of the future – which are rather bleak in this case – the respondent sees the future consequences of climate change as primary in relation to other threats to human society: climate change will result in more climate refugees, increased migration, food and water shortage, biodiversity loss and collapsing ecosystems, and struggles for the control of resources. This is a common topic in vernacular discourses on climate change, especially with regards to the topos of sinking islands in the Pacific (Ingemark 2019), though that connection is not brought up in this context. The temporality of apocalypse permeates much of the rest of the response, with other temporalities popping in and out of the discourse. One of these is the temporality of consumerism and the capitalist economy, which rests on an ever-increasing rate of consumption to uphold the economic system and provide all with a livelihood. The downside is that ecosystems crumble as a result, and the respondent states that she cannot fathom how these two things could possibly be reconciled.

She foresees the collapse and gradual demise of human society. Here two temporalities or timescales seem to co-exist: on the one hand, the sudden, catastrophic event of collapse and, on the other hand, the gradual or even lingering passing of human civilisation. This brings her to a predicted collapse of ecosystems, another sudden catastrophic event, and she wonders how we will find food for everyone under such deleterious circumstances. The theme of the disruption of ecological systems continues in her acerbic remark on our ways of speaking about other species as invasive, while we ourselves are just as invasive, swarming as we have the globe and using up all resources without any thought of tomorrow. She finishes off by landing in a negation of family time, otherwise a staple in much vernacular climate change discourse: family time is what we often use to make sense of climate change, since the temporal perspectives involved in climate change are quite difficult to grasp. Appealing to a better life for our children and grandchildren is an incentive to combat climate change (see Kverndokk 2017, 38). She concludes that her children will have a harsher life than she has had, and that she hopes they will opt to remain childless, breaking the link between the generations that have come before and the children who have no place in a catastrophic future.

It is easier to discern the diversity of temporalities and themes related to them in this text if we present it graphically. All of them involve very different timescales, timespans, cycles, rhythms, and paces, ranging from seasonal cycles, until now perceived as a more or less stable given in an otherwise changing world, to the disruptive temporalities of societal and natural collapse (Figure 4.1).

Yet despite this apparent multiplicity of temporalities, there is really only one that matters: the apocalyptic future of a climatically changed world. It swallows all other temporalities, sapping them of life. As Isak Winkel Holm has observed, climate change discourse often employs the prophetic mode, of which the primary characteristic is an ability to make us interpret the present in the light of an impending climate catastrophe (Chapter 6). The prophetic mode possesses a double temporality, as it encourages us to imaginatively place ourselves in the moment following the catastrophe and look back to our present, seeing the disaster as our fate (Dupuy 2013, 33). This fate can be one we can choose to avoid (Dupuy 2013, 33), but it can also, as in this response, appear utterly inevitable, making all resistance seem futile.

Since the prophetic mode is a combination of prospection and retrospection (Holm 2016, 93), giving us a stereoscopic vision of the present as a moment already passed, its peculiarity lies not in its revelation of an impossible future but in its simultaneous 'taking away' of the present. It robs us of 'any possibility of a firm, stable, lasting presence' (Blanchot 2003, 79; Holm 2016, 93–94). In this response to the questionnaire, and in many others, the effects of the prophetic mode become very evident; it deprives us of a firm, existential foundation in the present, and destroys the future.

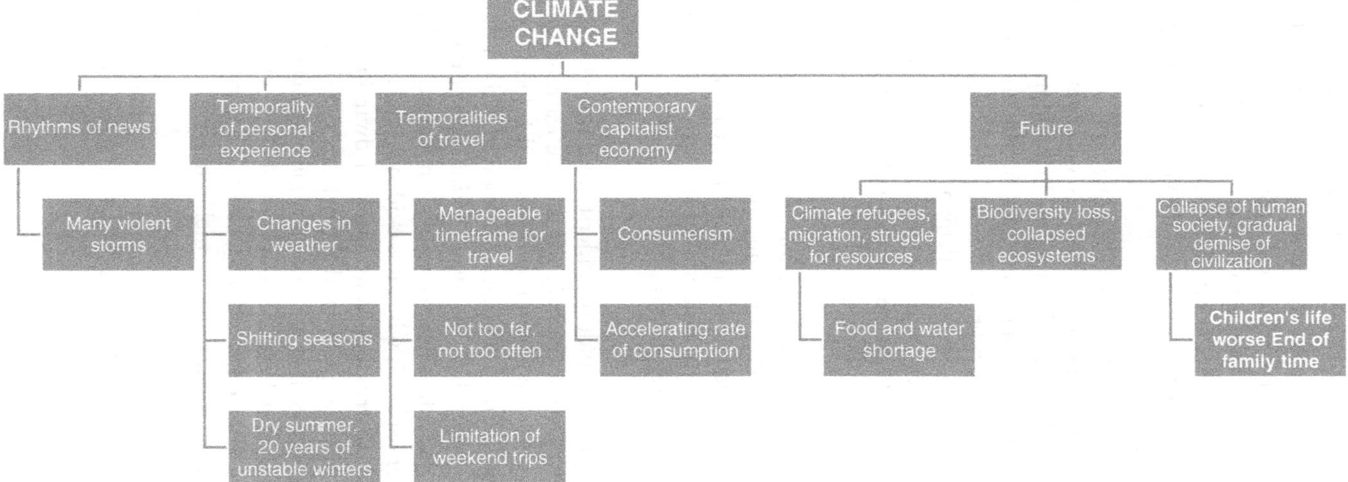

Figure 4.1 The temporalities in the response belong to five different but interrelated categories: the temporalities of news, of personal experience, of travel, of the capitalist economy, and of the apocalyptic future. The temporalities of news and travel are part of the current capitalist system and share its preoccupation with acceleration; in the case of travel, the respondent is consciously attempting to slow this down by choosing the more time-consuming practice of train travel rather than travelling by air. The temporalities of personal experience seem to function as prefigurations of the catastrophic future envisioned in the latter part of the response, where the capitalist economy is the cause of the destruction.

The future is bright

The next response is by a climate change denier responding to the Finnish questionnaire, and I have selected this one as it lends itself beautifully to the study of positioning and narrative voice. This reply is more complex in structure, as the respondent both has to defend himself against the implicit acknowledgement of the existence of climate change in the questions (despite apparent concessions to alternative views), and against an anticipated negative response to his own views.

In the response analysed above, news coverage was a recurrent point of reference. In this text, the respondent engages with news reports in such a way that it fundamentally shapes his replies. Ethnologist Oscar Pripp has called this phenomenon *the third interlocutor*, an interlocutor who is invisibly present in a discussion and induces people to answer questions that have never been posed (Pripp 2001, 71–75). Though Pripp was concerned with the face-to-face situation of an interview and the impact media reports had on the collection of his own material, the concept can also be fruitfully applied to responses to a written questionnaire, as it describes a similar process of relating replies to media output.

In the following quotation, the questions of the questionnaire – which were slightly different from the ones in the Norwegian and Icelandic questionnaires – have been included italicised in square brackets; replies to the third interlocutor are given in bold italics; and answers to an anticipated, more sympathetic addressee are in bold print:

> [***Third interlocutor***] It's difficult to understand how [why?] the mass media fuel these scares. Without having any reliable warrant for their claims. [*Do you think climate change is occurring right now?*] I don't believe in swift climate change, but I am afraid that a decrease in carbondioxide will affect plants needing it for their survival. [***Third interlocutor/*****Anticipated addressee**] Why make horror images of islands sinking [lit. drowning] as an effect of rising sea levels when it is evidently due to the "plate" sinking? [**Anticipated addressee]** Why speak of dwindling glaciers when they are growing in other places and the balance is satisfactory?
>
> [*Have humans influenced it or are there other causes of change?]* The greatest cause of snow and ice melting is the spread of small soot particles – not "greenhouse gases". [**Anticipated addressee**] All northern cultures have known for centuries that snow and ice melt when you sprinkle ashes, dust or something like this on the surface.
>
> [*What do you think the future will look like?*] The future is bright. [**Anticipated addressee?**] Plastic in the oceans, for instance, constitutes a real threat.
>
> (SLS 2303. Man)

As we can see, this respondent, a man of uncertain age, orients himself towards several audiences, beginning with the mass media which are chastised for churning out alarmist reports without apparent heed for the reliability and validity of their claims. Initiating the discussion by appealing to this third interlocutor – against which he positions himself in an antagonistic manner without having to openly confront it – he is setting the tone for the rest of the text. For by starting in such an adversative fashion, the brunt of his critique has to be borne not either by the persons who compiled the questions or by the anticipated addressee but by an undifferentiated collective entity. Perhaps the question was perceived as annoying, and this was a polite way of expressing displeasure without directly offending anyone.

The advantage of this rhetorical strategy is that it enables the respondent to reason with the compiler of the questions in a civil way while directing attention to issues he considers more pressing, such as the soot particles causing the melting of snow and ice in the Arctic. The 'horror images' of islands sinking into the sea probably refers to various island nations such as Tuvalu in Polynesia and the Maldives in the Indian Ocean, which gained international prominence in the wake of Al Gore's *An Inconvenient Truth* (2006). As in many other countries, these Pacific island nations have become iconic images of climate change in Finland where the respondent lives. Tuvalu in particular appears to function as a metonym for climate change in Western discourse, standing as part for the whole (Farbotko 2010; Ingemark 2019). The respondent outlines sea level rise as a purely natural process that is outside human control. This line of argument presupposes that he is accepting that the islands are actually sinking.

In the next sentence, the respondent addresses himself to an anticipated addressee, discussing the unfeasibility of speaking of dwindling glaciers. By employing a rhetorical question – 'Why speak of dwindling glaciers when they are growing in other places and the balance is satisfactory?' – he seems to include the addressee in a community of peers, who understand the improbability and vanity of doing so. Here, the respondent adopts an omniscient point of view, pronouncing the balance between ice sheets 'satisfactory'. This traditional narrative technique, a staple in Western literature (and film), implies that the narrator from whose point of view the events are perceived affects an omniscient, impartial perspective that is seemingly objective. Usually, this narrator speaks in third person, but here the respondent assumes this point of view speaking in first person. This switch of grammatical person somewhat undermines the authority of the statement, as it lays the respondent's interpretation open to criticism. The merit and success of the omniscient perspective from a narratological point of view relies on the 'impersonal' nature of the narration, which occludes the subjectivity of point of view; openly personalising it in this way somewhat diminishes its rhetorical efficacy.

Finally, turning to the compiler of the questionnaire, the respondent, quite contrary to many others, states that the future is actually bright, though this is immediately qualified by the very last sentence, which centres on the environmental degradation caused by plastic debris in the oceans. By juxtaposing two serious environmental concerns, the respondent offsets the issue of climate change by focussing our attention on another pressing problem.

To conclude the analysis of positioning, the orientation towards multiple addressees in this response results in a two-way addressivity: a backward-oriented one, to the questions put in the questionnaire and to the third interlocutor, and a forward-looking one, to an anticipated or imagined addressee. It is interesting to note that the respondent tackles the problem of potentially facing a reader negative to his views by addressing himself to a more favourably inclined, ideal or *implied reader*, one willing to accept his norms and values (for this term, see Booth 1983 [1961], 137–144). This manoeuver puts the unsympathetic reader at a disadvantage, since the refusal to accept the perspective of the implied author easily blurs into the moral failure or otherwise of the reader (Booth 1983 [1961), 138–139). Accordingly, we can assume that the compiler of the questions is not framed as the ultimate addressee of the response.

When it comes to temporalities, this response happens to start off in much the same way as the first one, with the temporalities of news coverage, which is the first kind of temporality appealed to in this text. The frequency of these alarmist reports seems to be an issue for the respondent, which implies a notion of pace and perhaps also acceleration. He does not believe in swift climate change but rather in longer processes of, implicitly, thousands or millions of years. Part of this long temporal perspective is the fear of a radical decrease of carbon dioxide in the atmosphere, eventually leading to the extinction of plants. While this is an ongoing natural geological process with a temporal scope of millions of years, cutting human emissions of carbon dioxide is perceived to exacerbate this process.

The 'horror image' of sinking islands once again appeals to the temporalities of news, which are juxtaposed to the longer natural processes of plate tectonics, which are favoured as an explanation for rising sea levels instead of the swift pace of climate change. The ratio of glaciers is then represented as a zero-sum game, with fluctuations over time in distribution and perhaps also density, but the amount remaining the same. Exactly how the temporality of this redistribution of glacial resources is envisioned is difficult to tell on the basis of the text, especially in its relation to the rapid melting people usually talk about. Is it constant but gradual, or does it occur suddenly? Since the respondent generally seems to privilege longer and slower temporal perspectives, the former might be more likely.

A longer temporal perspective is also implied in the discussion of black carbon emissions as a cause of snow and ice melt; this is framed as experiential knowledge that people in northern cultures have possessed for centuries – perhaps for millennia. Maybe this knowledge is also understood to have been passed on from generation to generation; this would make it an instance of family time, but with a much longer time perspective than the usual three-generation one. This stands in contrast to the rapid life cycle of the soot itself, which stays in the atmosphere for a few weeks at most (Novakov and Rosen 2013, 848).

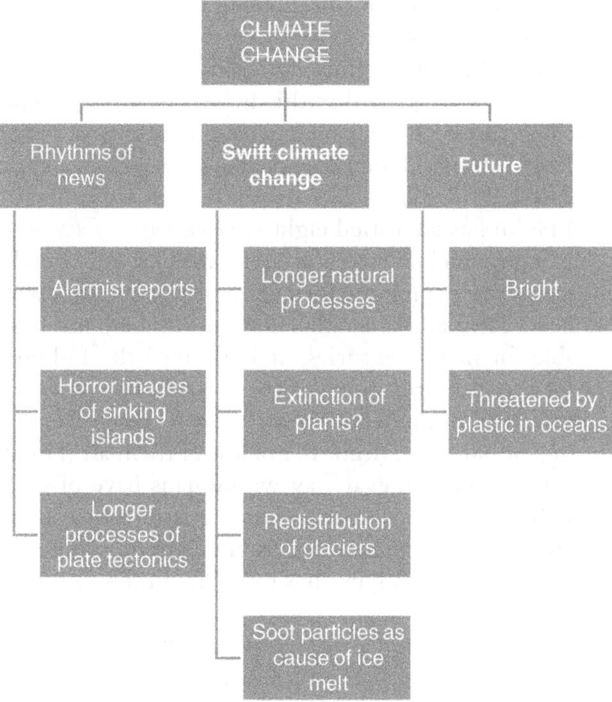

Figure 4.2 The temporalities of this response appear to rely on an opposition between longer temporal processes and swifter ones, in which the former are generally privileged. This means that climate change, as a relatively 'quick' process, is being denied in favour of longer natural processes, and actions taken to mitigate climate change are framed as potentially exacerbating the slow extinction of plants due to the gradual, naturally occurring loss of carbon dioxide in the Earth's atmosphere. The future plays a fairly small role as heavier emphasis is put on the distant past, but it is still presented with some apocalyptic overtones, given the threat of plastic in the oceans.

While the future is bright in this respondent's vision – the temporal span of which is imprecise – it is nevertheless a future that is threatened by something, namely plastic in the oceans. Even here, the future cannot stay totally unmarred by catastrophe, and this is quite remarkable; the apocalyptic pull seems too strong to resist entirely (Figure 4.2).

Climate change, catastrophe, and crisis

A significant share of the responses to the questionnaires adopted an apocalyptic view of the future. Thus, the first response discussed in this chapter was far from the only one envisioning the fall of human civilisation and the destruction of the world; it is a default mode of talking about the future. This apocalyptic framing appears to be so pervasive that it is even difficult to think of climate change without it (cf. Lilly 2016, 360). As stated in the beginning of this chapter, I am suggesting that this is because the concept of climate change is being organised by the overarching concepts of catastrophe and crisis, and that it has absorbed their age-old baggage of apocalyptic overtones (see also Kverndokk 2017, 33).

Isak Winkel Holm has identified eight *symbolic forms of disaster* that are the basic means through which Western societies make sense of and understand disaster. The age of these symbolic forms vary; some are old, whereas others are relatively new. They are also rather few in number, and have remained remarkably stable for many centuries, at least since the Lisbon earthquake in 1755. One of the most persistent of these symbolic forms is apocalypse, which presages the end of the world as we know it. Biblical as it is in origin, this is, of course, an old form, if not the oldest of them all (Holm 2012, 26); in a sense, it is the most primeval way we humans have of speaking about disaster. Personally, I think this exceptionally long history as part of our language of catastrophe might explain its perennial appeal; it might be difficult to separate the two, as if apocalypse has insinuated itself into the very DNA of disaster.

The modern development of the concept has probably contributed to its inextricability from climate change in common discourse. Since the Enlightenment, an intimate connection between catastrophe and human history has held sway, as catastrophe became a prerequisite for progress, civilisation, and rationality itself (Eliassen 2012, 45–47); or the unintentional result of our efforts to protect ourselves from, precisely, disaster (Rozario 2007, 13–14). In this respect, modern science can be viewed as part of the problem, as it has fostered an understanding of our planet as devoid of any intrinsic value. Climate change would then represent 'the catastrophic end' of this type of science (Lilly 2016, 361). At the same time, climate change has allowed contemporary scientists to adopt the apocalyptic rhetoric of the religion that science once abandoned, but with a secular basis of scientific measurements (Lilly 2016, 368–369). Many of the climate change deniers in my material object to this method of research communication, and to its dissemination in the media.

Kyrre Kverndokk has argued that the concept of crisis functions as a nodal point in climate change discourse; other concepts are organised by the concept of crisis and receive their meaning in relation to it. Climate catastrophe is one of these subordinated concepts, and it is the inverse of the notion of sustainable development. Thus, climate catastrophe and sustainable development form two divergent but nevertheless complementary future scenarios (Kverndokk 2017, 34). In this context, the exact hierarchy between the concepts of crisis and catastrophe is not so important. While they are also slightly different in nature as concepts, especially in temporal terms as catastrophe denotes a suddenly occurring event, a turning point, and crisis a more sustained temporal framework that includes a turning point and much else besides, there are some similarities between them, too. Today, the distinction has become less pronounced than it once was, and climate change is one of the reasons for this rapprochement between the concepts, as it merges the temporality of catastrophe with that of crisis (Holm 2012). Here, I am more interested in their common apocalyptic baggage, and both concepts appear to reinforce each other in the context of climate change, amplifying the apocalyptic propensities of one another.

Like the word catastrophe, crisis is a word of Classical Greek origin. It was primarily used in two different contexts, both of which are likely to affect our present-day understanding of the climate crisis. On the one hand, *krisis* referred to an act of judgement, often in a legal setting; with the advent of Christianity, this aspect of *krisis* was transposed to conceptions of the Last Judgement, where it denoted the judgement that would once and for all establish what true justice was (Koselleck 2006, 359). On the other hand, *krisis* was a medical concept. It could refer to the medical condition itself as well as to the judgement made about the course of a disease. In particular, it was linked to the turning point in the course of a disease, that moment in time when it would become evident whether the patient would recover or not (Koselleck 2006, 360). When the word crisis entered the vernacular languages of Europe, it was in this medical sense (Koselleck 2006, 362).

I suggest our contemporary notion of the climate crisis borrows from both the judicial and medical meanings of crisis. Anthropogenic climate change has made the Earth ill or even fatally ill, and this will lead to the end of the world as we know it. This idea connects to another development of the concept of crisis from the eighteenth century onwards, when crisis, just as catastrophe, became a template for interpreting history and historical time: history turned into one single crisis constantly taking place (Koselleck 2006, 371; cf. Kermode 2000, 93–98). This sense of constant crisis has become more acute with climate change.

The first respondent fully espouses this apocalyptic logic; in a sense, there is even a longing for the end, at least for the end of human civilisation. She regards it as a desirable development, while many others in the material speak of it as something to fear. As mentioned above, the second respondent, while otherwise resistant – if not allergic – to prophecies of doom, cannot wholly avoid introducing a sense of apocalyptic foreboding after all. In other responses, it is striking that the apocalyptic pull is so strong even in cases when respondents really want, and struggle, to maintain a positive outlook, as in the following reply:

> I like utopias, something to strive for, something we can dream of and hope for. My utopia is a society where humans have found their way back to their own place in nature. Where we have 'connected to' all that is living and growing and sprouting around us and where we succeed in

> seeing the consequences of all our actions. More concretely, I dream of more people, or as many as possible, beginning to grow their own food again, that we can teach our children to do the same. I dream of us ceasing to drive cars, ceasing to eat meat from animals in industrial agriculture, ceasing to hunt for status, power, and money. I also dream of a more close-knit society, where people see each other (this has, strictly speaking, nothing to do with climate change, but for me they are connected. If we respect the differences around us, whether it concerns people or nature, it will eventually impact the climate, too). But I'm also afraid that we will never get there. The world is, strictly speaking, heading in the wrong direction. Climate change has made it impossible to live in many places. Even greater waves of migration will occur than today, there will be more conflicts over land, water, and food. Climate change will probably alter the supply of these things in different ways.
>
> I think the world will be characterised by great waves of migration since large areas of land will not be arable any more. Naturally there will be more conflicts and [the world?] will be even more inhospitable than it already is. There will be starvation in the world and even greater differences between rich and poor. Another great challenge the world faces, which is both related to climate change and is independent of it, is the deterioration of the soil. Large parts of the world's arable land will not be fertile soon because it has been ruthlessly exploited since the advent of industrial agriculture. This will be a great, if not even greater threat to the global community than climate change is. That's what I think.
>
> (NEG 0263/00003. Woman)

Despite the respondent's partiality for utopias, apocalypse nevertheless encroaches on her vision of the future. She devotes slightly more space to writing about threats than she does to outlining a utopian future. Finding a new rapport and relation with nature is the basis for her ideal society, and this is quite in line with the recent evolution of the 'catastrophic logic of modernity' (Rozario 2007). Since nature is no longer a constant factor, progress must now be made on nature's conditions. As Kverndokk has observed, this means that nature – not progress – is considered the primary agent of change in history (Kverndokk 2017, 43).

This brings us back to deep time and the different variants of it espoused by those who think climate change exists and by climate change deniers. As I mentioned, the first group seldom speaks directly of deep time, and initially this puzzled me. Did they not have a sense of deep time? To be sure, it could be argued that it is possible to have an appreciation of climate change without involving deep time, since modern scientific measurements were introduced in the 1800s, and it is thus part of historical time rather than deep time. Yet it is difficult to imagine climate change and its impacts completely without the backdrop of a notion of deep time, since anthropogenic climate change is implicitly compared and contrasted to a standard of 'natural' climate change which implies a deep-time perspective. While the proxies of deep time, as outlined above, are robustly historical, they would not quite make sense without deep time. Conversely, some climate change deniers (or perhaps 'anthropogenic climate change deniers' would be a more appropriate designation since it is this particular aspect of climate change they uniformly oppose) do appeal explicitly to deep time, since they feel that they need to justify their points of view and therefore verbalise the deep-time basis for their line of reasoning.

One point of contention is, as hinted at above, the nature of deep time. Does evolution progress slowly and gradually, as Charles Darwin once claimed, or through sudden catastrophes, the most famous example of which is probably the Chicxulub bolide, whose impact wiped out the dinosaurs (Davies 2016, 31; Hessen 2018, 31)? The second respondent prefers gradualism, whereas the others are neocatastrophists. Another bone of contention is the character of the envisioned future: is it a future ruined by global warming and all its side effects, or are we approaching another ice age that we are now failing to prepare for since we spend our time worrying about anthropogenic climate change? As one climate change denier puts it:

> It has been much wilder and wetter in earlier periods than it is now. And I sincerely hope it will not turn as cold in the future as it might do; far, far more people are in danger of dying from the cold than from warming ....
> (NEG 0263/00091)

Climate change: a 'black hole' concept?

Returning to the proxies of deep time mentioned above, we might need to qualify Dipesh Chakrabarty's (2009, 201) contention that anthropogenic explanations for climate change have collapsed the distinction between natural history (deep time) and human history (historical time). While this is certainly true in a broad sense, it is important to note that they are not the only temporalities constituting the temporal dimensions of climate change. Many other temporalities are involved, some of them seemingly unconnected to the (collapsed) distinction between deep time and historical time. Since climate change is indeed capable of organising multiple temporalities, it affects the organisation of everyday life and media output as well, temporalities having little to do with notions of deep time or historical processes per se.

It is in this context that the concept of chronotopes becomes particularly useful, as it connects several chronotopic models under the umbrella term of climate change. Deep time and historical time are two of these chronotopic models, while lived experience and media temporalities constitute another two. Each has its own temporal logic, but all of them seem essential for making sense of climate change, whether people believe in its existence or not. Deep time and historical time are not enough to make sense of climate change beyond the intellectual level; if we want to understand it, in however imperfect a sense, and allow it to affect us and our ways of life, we need to integrate it emotionally and experientially into the fabric of everyday life. Media reports constitute one such means of integration, as they frequently appeal to our emotions and personal experiences in constructing their narratives.

This implies that sense-making is also inherently personal. Despite the fact that all respondents are drawing on shared chronotopes and chronotopic models, they still do so in their own ways and according to their individual preferences, presenting them in their own narrative voice to their imagined addressees.

★★★

These texts on climate change also invite more philosophical questions. Is climate change one of those 'black hole' concepts that suck in everything that come into their orbit? Or are we witnessing the beginnings of its transformation from a discourse to a figure of thought in philosopher Johan Asplund's sense – such as 'progress', 'childhood', or 'catastrophe' – the elemental constituents of our cultural thinking (Asplund 1979, 1991)? Asplund introduced the notion of figure of thought (Swedish: *tankefigur*) to label a level of historical analysis situated between the discursive one construed as part of a Marxist superstructure and the base (Asplund 1979, 149), but the pervasiveness of these figures of thought is what many commentators have fastened on. They are so inextricably linked to our culture and very way of life that it is nearly impossible to 'unthink' them without having to imagine a completely different society (Asplund 1979, 150–151). Thus, if climate change is gradually becoming a figure of thought, will it eventually engulf 'catastrophe' and 'crisis'?

The signs are there; in the material under study, it is quite common to allow climate change to swallow and sometimes blot out all other concerns, whether they are related or not. Such climate reductionism, for which the complex interactions of climate, environment, and society are reduced to merely climate (Hulme 2011, 247), abounds. This explains the ease with which climate change impacts on humanity, especially war, conflict, and migration, are predicted as necessary consequences. There is an epistemological comfort in the one-factor scenario: it is the one thing we can know about the future, when all else is uncertain (Hulme 2011, 247–249). The climate crisis and the climate catastrophe are also collective singulars: there is one all-encompassing crisis or catastrophe (Kverndokk 2017, 34). This way of speaking about them also contributes to the 'black hole' effect of climate change as a concept and phenomenon.

References

Archival material

Helsingfors, Finland:
Svenska litteratursällskapet i Finland, Folkkultursarkivet

SLS 2303

Oslo, Norway:
Norsk Etnologisk Gransking

NEG 0263

Printed

Agha, Asif. 2007. 'Recombinant Selves in Mass Mediated Spacetime'. *Language & Communication* 27: 320–335.

Asplund, Johan. 1979. *Teorier om framtiden*. Stockholm: Liber.

———. 1991. *Essä om Gemeinschaft och Gesellschaft*. Göteborg: Bokförlaget Korpen.

Bakhtin, Mikhail M. 1986a. *The Dialogic Imagination: Four Essays by M.M. Bakhtin*. Trans. Caryl Emerson and Michael Holquist, ed. Michael Holquist. Austin: University of Texas Press.

———. 1986b. *Speech Genres and Other Late Essays*. Trans. Vern W. McGee, ed. Caryl Emerson and Michael Holquist. Austin: University of Texas Press.

Bamberg, Michael. 1997. 'Positioning Between Structure and Performance'. *Journal of Narrative and Life History* 7 (1–2): 335–342.

Blanchot, Maurice. 2003. *The Book to Come*. Trans. Charlotte Mandel. Stanford, CA: Stanford University Press.

Booth, Wayne C. (1961) 1983. *The Rhetoric of Fiction*. 2nd ed. Chicago, IL and London: University of Chicago Press.

Chakrabarty, Dipesh. 2009. 'The Climate of History: Four Theses'. *Critical Inquiry* 35 (Winter): 197–222.

Davies, Jeremy. 2016. *The Birth of the Anthropocene*. Oakland: University of California Press.

Dupuy, Jean-Pierre. 2013. *The Mark of the Sacred*. Stanford, CA: Stanford University Press.

Eliassen, Knut Ove. 2012. 'Catastrophic Turns – From the Literary History of the Catastrophic'. In *The Cultural Life of Catastrophes and Crises*, edited by Carsten Meiner and Kristin Veel, 33–57. Berlin and Boston: De Gruyter.

Farbotko, Carol. 2010. 'Wishful Sinking: Disappearing Islands, Climate Refugees and Cosmopolitan Experimentation'. *Asia Pacific Viewpoint* 51 (1): 47–60.

Genette, Gérard. 1983. *Narrative Discourse: An Essay in Method*. Trans. Jane E. Lewin. Ithaca, NY: Cornell University Press.

Hessen, Dag O. 2018. 'Liv av død: Katastrofer som evolusjonær drivkraft'. In *Kollaps: På randen av fremtiden*, edited by Peter Bjerregaard and Kyrre Kverndokk, 31–45. Oslo: Dreyers forlag.

Holm, Isak Winkel. 2012. 'The Cultural Analysis of Disaster'. In *The Cultural Life of Catastrophes and Crises*, edited by Carsten Meiner and Kristin Veel, 15–32. Berlin and Boston: De Gruyter.

———. 2016. 'Under Water within Thirty Years: The Prophetic Mode in *True Detective*'. *Behemoth: A Journal on Civilisation* 9 (1): 90–107.

Hulme, Mike. 2011. 'Reducing the Future to Climate: A Story of Climate Determinism and Reductionism'. *Osiris* 26 (1): 245–266.

Ingemark, Camilla Asplund. 2019. 'Islands Submerged into the Sea: Islands in the Cultural Imaginary of Climate Change'. In *Former som formar: Musik, kulturarv, öar.*

Festskrift till Owe Ronström, edited by Camilla Asplund Ingemark, Carina Johansson, and Oscar Pripp, 199–208. Uppsala: Etnologiska avdelningen.

Kermode, Frank. 2000. *The Sense of an Ending: Studies in the Theory of Fiction with a New Epilogue*. Oxford and New York: Oxford University Press.

Koselleck, Reinhart. 2006. 'Crisis'. *Journal of the History of Ideas* 67 (2): 357–400.

Kverndokk, Kyrre. 2017. 'Klimakrisens tid'. *Arr Idéhistorisk tidsskrift* 2: 33–47.

Lilja, Agneta. 2016. '"Svara nu snällt på den lista, jag nu sänder!" – Om frågelistan som etnologisk arbetsmetod'. *Nätverket* 20: 20–27.

Lilly, Ingrid Esther. 2016. 'The Planet's Apocalypse: The Rhetoric of Climate Change'. In *Apocalypses in Context: Apocalyptic Currents through History*, edited by Kelly J. Murphy and Justin Jeffcoat Schedtler, 359–379. Minneapolis: Augsburg Fortress.

Marander-Eklund, Lena. 2016. '"Jag glömmer aldrig åskvädret 1960 tror jag det var" – ovädersberättelser'. *Svenska landsmål och svenskt folkliv* 139: 103–118.

Novakov, Tica and Hal Rosen. 2013. 'The Black Carbon Story: Early History and New Perspectives'. *Ambio* 42: 840–851.

OFV. 2019. *Bilåret 2018 – ett skritt nærmere 2025-målet*. Oslo: Opplysningsrådet for veitrafikken.

Prince, Gerald. 2003. *Dictionary of Narratology*. Rev. ed. Lincoln: University of Nebraska Press.

Pripp, Oscar. 2001. *Företagande i minoritet: Om etnicitet, strategier och resurser bland assyrier och syrianer i Södertälje*. Botkyrka: Mångkulturellt centrum.

Rozario, Kevin. 2007. *The Culture of Calamity: Disaster and the Making of Modern America*. Chicago, IL: University of Chicago Press.

Part 2

Mediating climate change temporality

5 The extreme summer of 2018

Norwegian weather news and the politics of weatherlore

Kyrre Kverndokk

Introduction

The world we live in is a weather-world, states anthropologist Tim Ingold. He argues that the weather is a fundamental human experience. People involve socially, affectively, and sensuously with the weather. Moreover, without weather – atmospheric pressure, liveable temperatures, wind, sun, and rain – life itself would be impossible (Ingold 2010, 132–133).

Heat is one such fundamental weather experience. The summer of 2018 was extremely hot and dry. There were record-breaking temperatures several places in North America, East Asia, and Europe. Heatwaves of this magnitude are slow and severe disasters. They do not only cause wildfires and failed crops. They also kill. Studies on mortality have shown that such heatwaves cause numerous deaths, especially among the elderly population (cf. Campbell et al. 2018). There is, however, a major geographical difference in what is defined as a heatwave. The term is relative and how it is used depends on the regional climate. In a country far to the north such as Norway, a heatwave is defined quite differently compared to southern Europe. The official Norwegian definition of a heatwave is three days with a maximum temperature above 28 degrees Celsius – in other words, quite pleasant temperatures (Berger et al. 2019). Thus, while the English term and most of its equivalents in European languages connote disastrous events, the Norwegian term does not necessarily have negative connotations. It is rather an ambivalent term that might reflect both desirable Norwegian summer temperatures and disastrous weather conditions, mediated through international news reports.

The long period of heat in Norway in 2018 started in early May and did not end until August. During this period, the average temperature was more than 4 degrees Celsius above normal.[1] July was also one of the driest months in modern Norwegian history.[2] The heat and drought were closely followed by media. This chapter is about how this heatwave was depicted in Norwegian news media.

Summer has traditionally been a season when the news supply drops in Norway, and political news is substituted for curiosities and everyday life events, as in other Western countries. In Norwegian, as in several northern

European languages, this period of time has been termed 'cucumber time'. The term refers to the fact that reports on the prices of cucumbers and strawberries find its way into news media. The weather is a favourite topic in both national and local news media during these weeks, and hot and sunny weather has usually been depicted in joyful terms, often illustrated by pictures of smiling sunbathers. The news coverage of the hot and dry weather in 2018 was, however, of a different kind. The local summer weather was seen in relation to the disastrous heatwaves elsewhere and was related to climate change. By means of this specific case, I want to discuss the relationship between immediate and large-scale, long-term perspectives in popular understandings of climate change.

In contrast to weather, climate change is calculatable and measurable but not directly observable. The long time frame and the global spatial range make climate change something rather abstract. This is pointed out by communication scholars as a major challenge for addressing climate change (e.g. Eide et al. 2010; Boykoff 2011; Eide and Kunelius 2012). In her book *Mediating Climate Change*, media scholar Julie Doyle claims that climate change must be '*made meaningful* in order to be able to better address this issue. How climate change is perceived – individually and collectively – depends upon how it is made socially and culturally meaningful to particular audiences' (Doyle 2011, 2, italics in the original text). She further argues that 'Climate change needs to be understood as a concern for the "here and now", rather than a distant future, "out there" somewhere. This involves making climate change temporally, spatially and socially meaningful and relevant' (Doyle 2011, 8). In this chapter, I will explore how climate change was turned into a 'concern for the "here and now"' through the way it was related to the 2018 heatwave in Norwegian news media. By close reading news articles on the heatwave, the chapter explores how the gap between the weather as a here-and-now experience and the global level of statistically measurable, anthropogenic climate change was bridged. Hence, in this chapter I examine how the weather experience of the hot and dry summer was made meaningful as an experience of climate change.

Weather, climate, news, and chronotopes

Ingold's approach to the weather-world is phenomenological and not necessarily compatible to a textual analysis. Yet, weather is also one of the most articulated parts of the human lifeworld. Everybody talks about it, and as sociologist Gary Alan Fine has pointed out, the weather is 'one of the shared interests that bind us together, building our social capital and contributing to the integration of civil society' (Fine 2007, ix). I regard such weather-talk as entextualisation of weather-worlds. Thus, in this chapter, I will approach the weather-world through the ways that human involvement with the weather is linguistically and textually presented.

The weather has often been regarded as a topic of vernacular conversation where people could have qualified opinions regardless of social hierarchies of expert systems (cf. Fine 2007). Climate change is quite the opposite. As it is not directly observable or experienceable, the identification of this phenomenon depends on highly qualified expertise capable of operating advanced computing and managing huge amounts of data. Popular understandings of climate change must therefore lean on this scientifically produced knowledge. Hence, relating local weather to climate change is not merely a matter of relating a particular case to statistical tendencies. It is also a matter of relating everyday life experiences to an abstract phenomenon made knowable through scientific expertise. Weather news is one genre where these two ways of knowing entangle.

The genre of news does not depend on any specific form or media. News might just as well take the shape of short notes, reports, or interviews, as of elaborate feature articles, and it might be communicated through print media, digital and visual media, or oral narratives. Rather than being defined through formal features, news might be defined by rhetorical characteristics underscoring the factuality, credibility, and objectivity of the content or what folklorist Elliott Oring has called the rhetoric of truth (cf. Oring 2012, 94–109). Such rhetoric is performed through references to trustworthy sources, a more or less plain language, and by reporting events in ways that make them appear likely to have happened. News is furthermore characterised by newness, newsworthiness, and immediacy. It is a genre depicting new information of interest for a public audience (cf. Stephens 2007, 6). News without immediacy is 'old news' or, in other words, no longer news. Chronotopically speaking, the timespace of news is first and foremost characterised by geographically located immediacy, by the portrayal of something that literally takes place.

News is the genre explored in this chapter, while *the chronotope* is used as an analytical device for examining how the tension between local weather experiences and the long-term process of anthropogenic climate change is handled. Literature scholar Mikhail Bakhtin regards the chronotope to be a structural element of genres (Bakhtin 1981, 84–85). Following Bakhtin, folklorist Camilla Asplund Ingemark argues that different generic chronotopes affect 'the way in which human beings and the world in which they live can be portrayed' (Ingemark 2016, 233). In other words, the chronotope of a genre such as news affects the way an event or phenomenon is depicted. Ingemark further remarks that the chronotope of a text 'is not only created through the fitting of the text to the generic model, but also by the specific intertexts – the other utterances – that are being absorbed and transformed' (Ingemark 2016, 234). The notion of the chronotope is, however, not limited to genres or texts; motifs and concepts may also have an embedded chronotopic structure. 'Climate change' is a chronotopically configured concept that has almost the opposite structure to the genre-specific chronotope of news due to its characteristically long-term temporality and large-scale spatiality. It appears in a range of genres, such as scientific articles, reports, political speeches, and news articles, to mention a few. Thus 'climate change' might also work as a device for 'dialogues of genres' (cf. Bauman 2004, 20), and might generate intergeneric and inter-chronotopic textual structures, but also chronotopic gaps and tensions (cf. Ingemark 2016, 249).

Furthermore, the chronotope, according to linguistic anthropologist Asif Agha, is not merely a generic representation of the interrelationship between time and space. A chronotope is more fundamentally also a semiotic representation of social practices, describing how a subject orients itself in time and space (Agha 2007, 321). Thus, it does not just have a spatial and a temporal dimension. It also has a social one, and different chronotopes organise the lifeworld of the subject in quite different ways.

The chronotope concept will be used in three different ways in the following analysis: as the timespace of textualised weather experiences, as the timespace of the notion of climate change, and as the generic timespace of the news genre.

Empirical material

The chapter will close read a limited number of news articles from both local and national media in Norway. The empirical material is based on searches in the media monitoring service Retriever, which covers all Norwegian newspapers[3] and media houses, in addition to some periodicals and press releases from larger organisations and think tanks. I have used search queries such as 'weather', 'climate change', and 'summer', and limited the searches to the period between mid-May and mid-August 2018. Most likely these queries do not cover all the published texts that relate the summer weather to climate change, but they undoubtedly give a good picture of the different ways it was done.

The search results cover a range of topics. During spring and the first half of summer, several local newspapers published articles that at the same time depicted the sunny, enjoyable weather and pointed towards worrying tendencies in a changing climate. From mid-July, this kind of article was accompanied by articles focussing on how the drought caused a severe wildfire hazard more or less all over the country and abroad, and how it also caused a national agriculture crisis due to crop failure. Additionally, a number of scientists, politicians, and environmentalists were interviewed or commented on the weather conditions through op-eds.

I have limited the empirical corpus to news articles that have the weather in Norway as the main topic, leaving out articles on agriculture and wildfires, but including interviews with politicians and scientists who commented on the weather situation. The material is further limited to newspaper articles. This limitation is based on the fact that newspapers still are widely read in Norway. They are important arenas for political and public debate, despite the last two decades of radically changing media technology. There were 223 newspapers in Norway by the end of 2017, with a total of two million printed copies and digital subscribers (Høst 2018, 5). The number equals approximately one copy for every second adult inhabitant of the country (Statistisk sentralbyrå 2020).[4]

The enjoyable heatwave as local summer weather

Among the newspapers that, already in May, were trying to cope with the gap between the wonderful local weather and global climate change was *Bergens Tidende.* On May 29, it published a two-sided article entitled 'The heatwave continues: There is now a queue of records'.[5] The article opens by stating that 'Yesterday's temperature of 29.8 degrees is currently the warmest temperature measured in Bergen in May, but now it is likely to get even warmer'[6] and continues:

> "The heatwave continues through Thursday and Friday, and no significant rainfall is expected. This afternoon we will get a light rainfall of 0.1 to 0.2 millimetres, and probably also on Thursday and Friday. Local afternoon showers will also fall in inland areas during the next few days. But that will not affect the forest fire hazard, which is very strong," says meteorologist on duty, Geir Ottar Fagerlid, at the Meteorological Institute in Bergen.
>
> Not only does the good weather continue, it is also possible that we get tropical nights this week, which means that the temperature does not fall below 20 degrees during the night' (Heian and Føli 2018, 6).[7]

Located between mountains on the western coast of Norway, the city of Bergen is known for its rainy climate. With an average of 2,380 millimetres of rain a year (Yr 2019), the dry and sunny weather was undoubtedly extraordinary. Thus, the small amount of rain forecasted by the meteorologist underscores the extraordinary nature of the weather conditions. Even though the high wildfire hazard is mentioned, such risk does not overshadow the impression of a typical feel-good story. The term 'the good weather' indicates a focus on joy and fun in the desirable sunshine. So is the mentioning of tropical nights. As in many European countries, a 'tropical night' is officially defined as a night where the temperature does not fall below 20 degrees Celsius. Such nights are rare in a coastal city like Bergen and are thus desirable.

The impression of a feel-good news story is underscored by the picture illustrating the article. It is an image of a lightly dressed man in a small boat. He is smiling and is obviously satisfied with the sunny weather and high temperatures. The caption reads, 'No reason to wear too much clothing these days' (Heian and Føli 2018, 6–7). Two side articles also underscore the impression of feel-good journalism. One of them reports good sales figures of ice cream, and the other reports that '[y]ou will now get a fine for barbecuing in the park'.[8] Yet, this latter information is also framed positively, focussing on how people are enjoying the good weather despite the prohibition of barbecuing, and it is illustrated by a picture of five lightly dressed young people happily relaxing in one of the parks of the city.

About halfway through the main article, the sub-headline 'climate change' indicates a shift of mode. The rest of the article is based on an interview with climate researcher Hans Olav Hygen, who is Director of Climate Services at the Meteorological Institute.[9] 'It is easy to enjoy real hot weather in Norway. But this is not only something positive', he states.

While the first half of the article portrays everyday life experiences of the local weather, the second half works to disseminate statistically based academic knowledge:

> Hygen emphasises that he is very careful about linking individual events to climate change. Initially, the spring heat is due to a high pressure that has been present for a long time. "At the same time, this falls into a pattern. In recent years, more frequent heat records have been set; it is quite clear that it is getting warmer".
>
> The average temperature of the globe has risen by 1.1 degrees since the end of the nineteenth century, mainly due to man-made climate emissions, according to the IPCC. In Norway, the temperature increase has been about 1 degree. Hygen says this underlying heating comes in addition to the effect of the high pressure over southern Norway. "A few decades ago, it probably wouldn't have been so hot in a similar situation".
>
> (Heian and Føli 2018, 6–7)[10]

This extract is characterised by a chronotopic shift. While the first part of the article is about the local here-and-now weather experience, following a typical chronotope of local news, the latter part is dominated by an almost opposite chronotopic model (cf. Agha 2007; Chapter 4). The pattern that Hygen refers to opens the scene in both spatial and temporal senses; so does the mentioned average temperatures. Such measurements make it possible to place the local weather event into global statistics with a time frame of about 150 years. However, the implied temporal framing is far wider. By using climate change as the frame, the local weather event is made relevant for the development of Earth history. Thus, the time frame is geological deep time, rather than just the last 150 years. This framing also makes the weather in Bergen at the end of May 2018 relevant for understanding future weather patterns – not in terms of the coming summer days, weeks, and months but in terms of the coming years, decades, and centuries. However, the immediacy embedded in the news genre typically centres around the present-day moment. Such immediacy also dominates this article. Even though the passage about rising global temperatures since the late nineteenth century explicitly refers to the past, the article in general is dominated by the use of present tense. That underscores how global climate change is presented as an ongoing contemporary process.

The contemporality of climate change is emphasised in the last paragraph of the article by referring to the weather situation far away from Bergen:

> Also Pakistan has had unusually hot weather lately. New heat records were set in April, and a week and a half ago, news came out that over 60 people had died of heat in the big city of Karachi – which, however, was denied by the local authorities.[11]

From a starting point with depictions of local leisure and pleasure, the focus of the text has changed to fatal consequences of a warmer global climate. With this information as a backdrop, the article ends with quite pessimistic future prospects for India and Pakistan:

> In some places in these countries, they are approaching a limit that might make it impossible to live there over time. "There are well over 1 billion people living there, and in some areas, we must expect people to migrate," says the researcher [Hygen].
>
> (Heian and Føli 2018, 7)[12]

In this article, typical feel-good summer news is paired with a dissemination of climate science, which gives the local weather global significance. Through the locally experienced weather, *Bergens Tidende* is pointing towards a dystopic future, not necessarily locally but on a global scale. The way this is done is quite typical for how news media relate extreme or abnormal weather events to climate change (Kverndokk 2019). Statistics and scientific references enable a bridging of the chronotopic gap between the here-and-now of the local news and large-scale timespace of the climate change concept. Spatially speaking, local weather is made globally relevant, while temporally speaking, the dominant use of present tense privileges the present-day events in the description of both past and future climate.

In the two following sections, I will discuss two implications of the way that the local weather event is made globally significant through bridging this chronotopic gap. I will first explore how such bridging may influence how weather experiences are individually perceived, and in the last section of the analysis I will discuss how it also facilitates the politicisation of weather experiences.

An affective moment of a changing climate

In mid-July, after the heatwave had lasted for about two months, the newspaper *Nordlys*, located in Tromsø in northern Norway, published an article about quite an extraordinary weather phenomenon. Under the headline 'I was fascinated and almost a bit scared',[13] a vacation visitor, Bjørn Salvesen, was interviewed. He had experienced a sudden temperature rise when he was walking along the seashore just outside of Tromsø in the middle of the night.

In just ten minutes the temperature rose from 18 to 26.3 degrees Celsius. 'We were out walking along the shoreline around 1 a.m., when the wind began to blow a little bit and we saw dark clouds', he explains and continues, 'We expected it to get cooler, but that didn't happen at all. Within minutes, it was tropically hot, in the middle of the night out here at the shore of the open ocean' (Nielsen 2018).[14] The assumption that sudden wind and dark clouds indicate a shift to a colder weather is a typical experience-based vernacular knowledge. The surprise when the total opposite happened is underscored by his careful reporting of certain details when he is describing the event, such as twice mentioning the time of the day, notifying the short time frame, emphasising that it happened 'out here at the shore of the open ocean', and by using the term 'tropical heat' (Nielsen 2018). Just like 'tropical nights', 'tropical heat' indicates quite extraordinary warm temperatures. Such heat, however, is not really tropical in a strict sense. The term is commonly used positively in Norwegian, often describing desirable summer temperatures. Yet, in this short description from a place far north of the Arctic Circle, it is used ambiguously. 'This is not supposed to happen',[15] he concludes. The article further quotes a Facebook post about the event, in which Salvesen wrote: 'This is seriously something to worry about'.[16]

His description demonstrates how people engage tactilely and affectively with the weather, as living beings in a weather-world (cf. Ingold 2007; 2010). In line with Ingold, geographer Mike Hulme emphasises the importance of understanding the experience of weather and climate as a starting point for discussing what he terms 'cultures of climate' (Hulme 2017). Hulme points out that while the weather is uncontrollable and constantly shifting, the notion of climate has traditionally represented stability and predictability. And, while the weather might be dramatic, unpredictable, and chaotic, climate has been something to put trust in. '[The climate] offers a way of navigating between the human experience of a constantly changing atmosphere, with its attendant insecurities, and the need to live with a promise of stability and regularity', he states (Hulme 2017, 5). In that regard, both weather and expectations of the local climate involve affects and emotions.

This is also true of Salvesen's account of his experience. He depicts a complex affective relationship between the experienced weather, local climate, and global climate change. He seems to be well aware of how climate research is based on statistics and probability and says to the newspaper, 'Even though one should be careful not to link single phenomenon to climate change, I have *certainly* been thinking a little bit about this' (my italics).[17] The adverb *certainly* implies an underlying worry about the changing climate. It indicates that the awareness of a changing climate is framing his experience, not so much intellectually as affectively. The way he frames his weather experience with an awareness of global, climatic processes might be described as a kind of late modern weatherlore. As such, the newspaper article does more than simply depict a quaint experience. It also describes how to deal with the fundamental uncertainty of changing climate and changing weather patterns.

One important aspect of weatherlore has traditionally been risk perception. Reading the weather and signs in nature to forecast the coming weather has been a way of giving some sort of predictability to the unruly weather. In a world before modern meteorological forecasting was available through live broadcasting, vernacular forecasting on a daily basis was crucial for evaluating risks for occupations that rely on the weather, such as fishing, farming, shepherding, and logging (cf. Hodne 1994; Fulsås 2003). Long-term seasonal weatherlore was also a kind of risk perception in a traditional agrarian society. To read signs in nature during winter and spring to predict the coming summer weather was a way of predicting the harvest (Bondevik 1950; Hodne 1994, 10–12). The rich tradition of weatherlore is complex, drawing on both beliefs and empirical observations (Wurtele 1971, 923). Yet, the stability of the local climate has been an important premise for this tradition, enabling vernacular predictions to draw on accumulated weather experiences that had been transmitted through multiple generations. In that sense, traditional weatherlore is based on a local timespace, characterised by stability over time.

During the last three to four decades, the relationship between weather and climate has been turned upside down. While upcoming weather is now quite predictable due to 24/7 online weather services, the climate is changing into a new and uncertain state. Global warming has added a fundamental uncertainty to the notion of climate. Even though the future climate is being modelled scientifically, such model simulations are to some extent uncertain, especially when it comes to predicting local and regional climates. Even more importantly, a changing climate will certainly also affect weather patterns, and weather extremes will occur more frequently and at unexpected times. These extremes will be of a kind that might not have been experienced before. This development changes the perception of both climate and weather. Weatherlore has traditionally been based on accumulated experiences, transmitted vernacular knowledge, and beliefs, while weatherlore in the age of climate change is framed scientifically. This especially counts for long-term weatherlore. Such weatherlore is not particularly occupied with predicting upcoming seasonal changes but instead tries to understand what awaits us in the years and decades to come. One example of such weatherlore is the way popular media often considers extreme weather events to be caused by a changing climate, while such events at the same time also prefigure severe future changes in climate (Kverndokk 2019). This also seems to be the case with Salvesen's experience that summer night in 2018. When his short-term, local meteorological prediction failed, he re-interpreted his experience in large-scale climatic terms. The scary and unusual weather event is perceived as a local unfolding of the changing climate, and also as pointing towards a climate-changed future.

Salvesen's experience has an affective dimension beyond the immediate. The weather is no longer just a local concern; it also takes on a long-term global dimension. This is reflected in his comment that '[t]his is seriously something to worry about'. He expresses a notion of fear and uncertainty of the unpredictable and potentially disastrous future climate beyond the actual place where he experienced the sudden shift in temperature. This kind of long-term weatherlore is a vernacular version of what sociologist Ulrich Beck terms 'the semantics of risk' that is 'the present thematization of future threats' (Beck 2009, 4). Such semantics involve both professional and lay practices of identifying and describing risks. The professional practices are characterised by numerical calculations and scientific modelling, while the vernacular practices use an awareness of these calculations as an interpretive key for reading weather as communicative signs that reveal future risks (cf. Kverndokk 2019, 315–316).

Thus, weatherlore in the age of climate change is founded on a chronotropic timespace that is global and implies radical changes over time. A weather experience is, however, necessarily local and immediate. The orientation in space and time in Salvesen's story is at the same time immediate and large-scale. This duality in terms of timespace makes the experience fundamentally ambiguous. It is both scary and desirable, with its pleasant summer temperatures: 'I could not go to bed, I had to enjoy it', he told the newspaper. The entire article is characterised by the ambiguity between enjoying the moment and worrying about the future.

The politicisation of the summer weather

In late June, when the heatwave had lasted for about a month, Prime Minister Erna Solberg, representing the Conservative Party,[18] was asked by *Stavanger Aftenblad* to comment on Hygen's claim (discussed above) that the weather situation indicated a changing climate. The reporter asked, 'Are you primarily happy or worried about the abnormally high summer temperatures – and do you think this is a symptom of man-made climate change?' (Søndeland 2018).[19] Her response was:

> I am very careful to stress that we must not mix up climate and weather. These are two different things. Climate change is something that happens over a long period of time. The weather varies a lot from year to year.[20]

And she continued:

> If you had asked me this question in February, I could have given a completely different answer than now. Because we have had a cold and nice winter. There are weather differences. Then it might be that the nice weather is related to climate, but we must have a long time frame to be sure. (Søndeland 2018)[21]

According to the newspaper, this reply was simply a short popular lecture about climate change. Her distinction between weather and climate certainly draws on climate research. However, the answer does far more than just enlighten the general public. She is careful not to relate the weather situation to climate politics. This is done elegantly in two different ways. First, by referring to the scholarly informed distinction between weather and climate. Second, she elaborates on this distinction by referring to the cold winter the same year. By using the adjective *nice* about both the cold winter and the warm and sunny summer, she appeals to cultural notions of seasons. The northern European winter is supposed to be cold, while in contrast, the summer weather is supposed to be warm. In Norwegian, the adjectives Solberg uses, *cold* and *nice*, often appear in a pair when describing an ideal winter, while *nice weather* during the summer months implies high temperatures and sunny days. By using such cultural markers, she normalises the weather situation as not just harmless but also desirable and enjoyable. Thus, she does not just enlighten the public, she also encourages people to just enjoy the weather.

While Solberg carefully avoids framing the weather politically, her Minister of Climate and Environment did the exact opposite a month later. In the meanwhile, the consequences of the dry, hot weather had escalated. It had been a growing concern for crop failure in Norwegian agriculture. By mid-July, smaller or larger forest fires appeared frequently, and news media also started to report on forest fires out of control in Sweden. The language in these reports had a distinctly apocalyptic tone (e.g. Kvistad 2018). This was the backdrop for an interview the national newspaper *Aftenposten* did with the Minister of Climate and Environment, Ola Elvestuen, on July 28. Under the headline 'More people are now realizing the seriousness', he says:

> "I think many people feel that the summer has not been quite the way it should be" […] He himself is one of them. Elvestuen does not doubt that the last weeks' intense heatwave is a result of enduring climate change and global warming.[22]

And he concludes, 'We are right in the middle of climate change that is facilitating for a more intense weather situation' (Mauren 2018).[23]

The Norwegian government was at that time a coalition of three conservative and liberal parties. Elvestuen, representing the Liberal Party,[24] is, just like Solberg, drawing on climate research. However, in line with several of the climate researchers that engaged in the public debate that summer, he does not merely separate between weather and climate. He also underscores the consequences of shifting baselines (e.g. Joner and Zondag 2018). In doing so, he, in contrast to Solberg, moves the weather to the centre of environmental politics. He uses the weather situation to support 'calls for a new mobilization to reach the world's climate targets'.

There are striking similarities between Salvesen's depiction of his experience and how Elvestuen describes the weather. They are both referring to the weather in affective terms as something that 'signal[s] more than a "nice summer"', as Elvestuen puts it (Mauren 2018).[25] Elvestuen describes the weather as something that feels fundamentally wrong, and he uses his own experience as a starting point to make a general claim: 'I think many people *feel* that the summer has not been quite the way *it should be*' (my italics). The assumed general feeling is framed by scientifically informed knowledge. This framing enables the experience to be textualised, not merely by putting it into words but also by turning it into readable signs pointing towards global climatic changes and a gloomy future. This kind of semiotic practice is, as in Salvesen's case, a kind of long-term weatherlore in the interface between the dissemination of statistics and modelling, and vernacular reasoning. It draws on the same complex chronotopic construct as in Salvesen's case, being at the same time local, immediate, and enjoyable as well as global, future-oriented, and worrying. This interpretive practice is, however, not about the future as such; it is rather about the relationship between the present and the future. The future is displayed in the present weather as a warning about *what might happen* if measures are not taken. Elvestuen demonstrates how the semantics of weather is not just about naming risks, it is also about identifying required measures. It is as much about politics in the present than it is about the future. In Elvestuen's case, it is used as an argument for fulfilling the tasks outlined in the Paris Agreement.

The interview was followed up by both interviews and op-eds where politicians across the political spectrum commented on the relationship between the weather and climate politics. Among the first to speak was the environmentalist and parliamentary member for the Socialist Left Party,[26] Lars Haltbrekken. He criticised the liberal-conservative government for not doing enough to cut the emissions of greenhouse gasses. Using the heatwave as his starting point, he claimed that Elvestuen's statement was 'just talk'.[27] 'The government has not only been incapable of taking measures. They have also been advocating for increasing the emissions',[28] he said, with a reference to the planned opening of a large new oilfield which is scheduled to run for 40 years (Mauren and Riaz 2018).[29] In the same article, Kari K. Kjos, representing the right-wing Progress Party[30] in the parliament, took the opposite stand. She claimed, 'I am not sure if the heat comes from climate emissions. Because the summer last year was characterised by a lot of precipitation. I think we are lucky to have got such a nice summer this year' (Mauren and Riaz 2018).[31] Her separation of politics and personal weather experiences was in line with what Prime Minister Solberg said a month earlier. Yet, the period of drought and the frequent wildfires had in the meanwhile made it clear that such a statement was far from apolitical.

Interviews with Haltbrekken and Kjos were framed by the header 'The Extreme Heat', explicitly inscribing the heatwave in the sphere of extreme weather, often associated with climate change (cf. Tvinnereim and Fløttum 2015). The interviews were also published side by side with an article listing extreme weather events from around the world, such as exceptionally high temperatures in the Middle East, North Africa, and Canada; wildfires in Sweden, Greece, and California; drought in Russia, France, and Germany;

floods in Japan; and the high temperatures in Norway. This listing was discussed in general terms by two climate researchers and illustrated by a photograph of the forest fires in Sweden (Christiansen and Fjelltveit 2018). In this way, the newspaper left no doubt that the heatwave in Norway was part of a worldwide pattern of extreme weather events related to climate change.

This framing made Kjos's statement look like denial of observable facts. Her claim that 'we are lucky to have got such a nice summer this year', seems at first glance to be an everyday apolitical reflection. Yet, the way the newspaper framed her statement made it obvious that it was a political statement. When she declared that she was 'not sure if the heat is due to climate emissions', and pointed towards annual variations in weather, she was paraphrasing the official programme of the Progress Party. The party advocates climate politics based on a precautionary principle, acknowledging global changes in climate but claiming that 'we know too little about what influences these changes' (Fremskrittspartiet 2017–2021).[32] The programme further emphasises that 'it might be unfortunate and incorrect to link any flood, heat or cold wave, storm, and other weather forms to a claim about man-made climate change' (Fremskrittspartiet 2017–2021).[33] This might be a way of taking a stand against environmentalism since it has become quite common among both environmentalists and in popular media to use extreme weather events to argue for efficient climate politics (cf. Kverndokk 2019). The party's approach to climate change balances at the edge of climate scepticism, and some leading members are self-declared sceptics, including the former leader of the party. Kjos's concluding remark to the reporter explicitly followed this sceptical approach. She stated, 'there is nothing new about weather variations. We are quite sceptical of the theory of man-made climate change, but we are concerned about safeguarding the environment' (Mauren and Riaz 2018).[34] Her insistence of separating the weather from politics is based on a reluctance to institute regulation of and restrictions on the consumption of fossil fuels, car driving, and other aspects of everyday life. Chronotopically speaking, she insisted on maintaining the gap between the here-and-now of the local weather experience and the large-scale timespace of the notion of climate change.

In light of how the relationship between weather and climate is described in the Progress Party's programme, it is also tempting to understand Prime Minister Solberg's statement as a way of carefully manoeuvring away from a delicate political disagreement between the coalition parties in her government. Her statement is at once satisfying the Progress Party's attempt to separate summer life from political restrictions, and still acknowledging climate change. It is an elegant political balancing act in a time where the weather has turned into a political issue.

The process of transforming the weather into a political issue has been going on for several years. During the summer of 2018, it became clear that it was no longer just storms and dramatic weather events that were related to climate change. The kind of weather that traditionally has been a subject for feel-good news stories had also become politicised. The news coverage of the summer weather of 2018 demonstrated that the weather had become politicised to such an extent that it seems to have been almost impossible – at least for a politician – to just enjoy it, without making a political statement. This politicisation of the weather was chronotopically organised. It was a matter of bridging the gap between the timespace of local weather news and weather experiences and the timespace embedded in the notion of climate change, or a matter of keeping this gap open by differentiating between the local weather and global climatic changes.

Conclusion

This chapter is a three-step analysis of how the local summer weather of 2018 was used as a way of making climate change newsworthy as a contemporary event. The analysis of the article from *Bergens Tidende* demonstrated how the conceptual chronotope of 'climate change' was put in dialogue with the here-and-now chronotopic structure of weather news, making the weather at once both locally and globally significant. Temporally speaking, long-term and large-scale anthropogenic climate change was focalised in the present moment through the weather. This double chronotopoi is fundamentally ambiguous. It makes the weather at once both enjoyable and alarming.

The chapter has especially explored how this chronotropoi enabled a generic shift from feel-good news to weatherlore. In a traditional agrarian society, weatherlore was not just practice for naming risks, it was also a way of regulating social behaviour. For instance, seasonal weatherlore predicted the coming seasonal weather and thus also regulated agricultural activities, such as when to sow and harvest, and when to let cattle out on the summer mountain pastures, which in Norway was communal grazing ground (cf. Bondevik 1950; Solheim 1952, 91–98). In that regard, weatherlore was a way of making sure than no one would benefit from sowing or harvesting at another time than the rest of the community or benefit from having their cattle graze in the mountains and hills before anybody else. Late modern weatherlore of the kind I have discussed in this chapter also involves the regulation of social life, by serving as the basis for arguments in favour of regulating the consumption of fossil fuels. Such weatherlore is a means for understanding what might lie ahead if the emissions of greenhouse gasses do not radically decline. In that regard, it is fundamentally political. It is an interpretive practice unfolding in an interface between science, politics, and everyday life. As the object of such an interpretative practice, the weather is no longer just 'one of the shared interests that bind us together, building our social capital and contributing to the integration of civil society' (Fine 2007, ix). It is also quite the opposite – a potential political battlefield.

In other words, to connect locally experienced weather to climate change and the prospect of a climate-changed future is more than a matter of scaling down climate change into something immediate and observable. It is also a matter of transforming abstract, scientifically produced knowledge and geopolitical issues into the affective and sensory experiences of a changed and still changing climate. Thus, it is not just the weather that is ascribed a political dimension through such weatherlore but also the weather-world. The politics of climate change has also become the politics of vernacular weather experiences.

Notes

1 With climate averages 1961–1990 as the reference value.

2 Only July 1901 has been recorded as a month with less rainfall on a national scale, Grinde et al. 2018, 4.

3 There is one exception. The newspaper *Dagens næringsliv* withdrew from Retriever in 2017.

4 The total population over eighteen years old in Norway is 4,100,000 (2017) according to Statistisk sentralbyrå 2020.

5 'Hetebølgen fortsetter: Nå star rekordene i kø'.

6 'Gårsdagens temperatur på 29,8 grader er foreløpig den varmeste temperaturen som er målt i Bergen i mai, men nå blir det trolig enda varmere'.

7 'Hetebølgen fortsetter til og med torsdag og fredag, og det er ikke ventet nedbør av betydning. I ettermiddag får vi et lite drypp på 0,1 til 0,2 millimeter, og sannsynligvis torsdag og fredag også. Lokale ettermiddagsbyger blir det også i indre strøk de nærmeste dagene. Men det vil ikke gjøre noe med skogbrannfaren som er meget stor, sier vakthavende meteorolog Geir Ottar Fagerlid ved Meteorologisk institutt i Bergen.

Ikke nok med at godværet fortsetter, men det ligger også an til at vi får tropenatt denne uken, altså at temperaturen ikke går under 20 grader i løpet av natten'.

8 'Nå får du en bot for å grille i parken'.

9 The interview with Hygen was even distrusted by the Norwegian News Agency and published as an independent interview or in relation to local weather at other locations.

10 'Det er lett å glede seg over skikkelig varmt vær i Norge. Men dette er ikke bare positivt, sier Hygen, som er klimaforsker og avdelingssjef ved Meteorologisk institutt.

Hygen understreker at han er svært forsiktig med å knytte enkelthendelser til klimaendringene. I utgangspunktet skyldes vårvarmen et høytrykk som har blitt liggende lenge.

–Samtidig føyer dette seg inn i et mønster. I seinere år er det satt hyppigere varmerekorder, det er helt tydelig at det blir varmere.

Gjennomsnittstemperaturen på kloden har steget 1,1 grad siden slutten av 1800-tallet, hovedsakelig på grunn av menneskeskapte klimautslipp, ifølge FNs klimapanel. I Norge har temperaturøkningen vært omtrent 1 grad. Hygen sier denne bakenforliggende oppvarmingen kommer i tillegg til effekten av høytrykket som ligger over Sør- Norge.

– For noen tiår siden ville det trolig ikke blitt fullt så varmt i samme situasjon'.

11 'Også Pakistan har hatt uvanlig varmt vær den siste tiden. I april ble det satt nye varmerekorder, og for halvannen uke siden kom det meldinger om at over 60 mennesker hadde dødd av hete i storbyen Karachi – noe som imidlertid ble benektet av de lokale myndighetene'.

12 'Noen steder i disse landene nærmer de seg en grense som kan gjøre det umulig å bo der over tid.

–Det bor godt over 1 milliard mennesker der, og i noen områder må vi forvente at folk kommer til å flytte på seg, sier forskeren'.

13 'Jeg ble fascinert og nesten litt redd'.
14 'Vi var ute og gikk i fjæra ved 0100-tida da det begynte å blåse litt og vi så mørke skyer. Vi forventet at det skulle bli kjøligere, men det ble det slett ikke. I løpet av få minutter ble det tropisk varmt, midt på natta her ute i havgapet'.
15 'Det skal liksom ikke skje'.
16 'Dette er seriøst bekymringsfullt'.
17 'Selv om man skal være varsom med å koble enkeltfenomener opp mot klimaendring, så har jeg jo tenkt litt på dette'.
18 Høyre.
19 'Er du først og fremst glad eller bekymret over de unormalt høye sommertemperaturene – og tror du dette er et symptom på menneskeskapte klimaendringer?'
20 '– Jeg er veldig opptatt av at vi ikke må blande sammen klima og vær. Det er to forskjellige ting. Klimaendringer er noe som skjer over lang tid. Været har stor variasjon fra år til år'.
21 'Hadde du spurt meg dette spørsmålet i februar, kunne jeg gitt et helt annet svar enn nå. For vi har hatt en kald og fin vinter. Det er værforskjeller. Så kan det være at det fine været henger sammen med klima, men det må vi se over lengre tid'.
22 - Jeg tror mange opplever at sommeren ikke har vært helt slik som den burde være [...]. Han er selv en av dem. Elvestuen er ikke tvil om at den intense varmebølgen de siste ukene er et resultat av varige klimaendringer og global oppvarming.
23 – Vi står midt oppe i klimaendringer som er med på å gjøre værsituasjonen mer intens'.
24 Venstre.
25 'signaliserer mer enn en "fin sommer"'.
26 Sosialistisk venstreparti.
27 'mest prat'.
28 'Regjerningen er ikke bare handlingslammet. De har også gått inn for nye utslippsøkninger'.
29 He refers to the Johan Castberg oil and gas field in the Barents Sea, which is planned to open in 2022. It is the largest planned oil and gas field in the world. It is much disputed, not merely because of its size but also due to its location in vulnerable natural environments in the far north.
30 Fremskrittspartiet.
31 '– Jeg er usikker på om varmen kommer av klimautslipp. For sommeren i fjor var preget av mye nedbør. Jeg synes vi er heldige som har fått en så fin sommer i år'.
32 'om hva som påvirker disse endringene'.
33 '[D]et [kan] være uheldig og uriktig å koble enhver flom, hete- eller kuldebølge, storm og andre værformer til påstanden om menneskeskapte klimaendringer'.
34 '[D]et er ikke noe nytt at været varierer. Vi er veldig skeptiske til teorien om at klimaendringer er menneskeskapte, men vi er opptatt av å ta vare på miljøet, sier Kjos'.

References

Agha, Asif. 2007. 'Recombinant Selves in Mass Mediated Spacetime'. *Language & Communication* 27: 320–335.

Bakhtin, Mikhail. M. 1981. *The Dialogical Imagination. Four Essays by M.M. Bakhtin*. Ed. Michael Holquist. Austin: University of Texas Press.

Bauman, Richard. 2004. *A World of Others' Words: Cross-Cultural Perspectives on Intertextuality*. Oxford: Blackwell Publishing.

Beck, Ulrich. 2009. *World at Risk*. London: Polity Press.

Berger, Anniken Celine, and Amalie Kvame Holm. 2019. 'Stadig flere hetebølger i Norge'. *Meteorologisk institutt*. https://www.met.no/nyhetsarkiv/stadig-flere-hetebolger-i-norge, accessed December 19, 2019.

Bondevik, Kjell. 1950. *Jordbruket i norsk folketru: Ei jamførande gransking*. Vol. 2, Langtidsvarsel. NFL 66. Oslo: Norsk Folkeminnelag.

Boykoff, Maxwell. 2011. *Who Speaks for the Climate? Making Sense of Media Reporting on Climate Change*. Cambridge: Cambridge University Press.

Campbell, Sharon, Tomas A. Remenyi, Christopher J. White, and Fay H. Johnston. 2018. 'Heatwave and Health Impact Research: A Global Review'. *Health & Place* 53 (September): 210–218.

Christiansen, Hanne, and Ingvild Fjelltveit. 2018. 'Slik har ekstremværetpreget verden'. *Aftenposten*, July 30, 16–17.

Doyle, Julie. 2011. *Mediating Climate Change*. Farnham: Ashgate.

Eide, Elisabeth, and Risto Kunelius, eds. 2012. *Media Meets Climate. The Global Challenge for Journalism*. Göteborg: Nordicom.

Eide, Elisabeth et al., eds. 2010. *Global Climate, Local Journalisms: A Transnational Study of How Media Make Sense of Climate Summits*. Bochum: Projectverlag.

Fine, Gary Alan. 2007. *Authors of the Storm: Meteorologists and the Culture of Prediction*. Chicago, IL: Chicago University Press.

Fremskrittspartiet, Klima. 'Partiprogram 2017–2021'. https://www.frp.no/tema/miljo/klima, accessed January 9, 2020.

Fulsås, Narve. 2003. *Havet, døden og vêret: Kulturell modernisering i Kyst-Noreg 1850–1950*. Oslo: Samlaget.

Grinde, Lars, Elin Lundstad, Reidun Skaland, and Helga Therese Tilley Tajet. 2018. *Været i Norge: Klimatologisk månedsoversikt*. Oslo: Meteorologisk institutt. https://www.met.no/publikasjoner/met-info/met-info-2018/_/attachment/download/36d3a6b7-0d67-4b5b-ae90-811bf2ff0256:13d3bbc6029bb41d73fa96e-8ba0b00c26533428f/MET-info-07-2018.pdf, accessed June 12, 2019.

Heian, Hilde and Are Føli. 2018. 'Varmebølgen fortsetter: Nå står rekordene i kø'. *Bergens tidende*, May 29, 6–7.

Hodne, Ørnulf. 1994. *Gamle norske værvarsler. 1700 værtegn fra hele landet*. Oslo: Cappelen.

Høst, Sigurd. 2018. *Avisåret 2017*. HVO; Rapport 86/2018. Volda: Høgskulen i Volda. https://bravo.hivolda.no/hivolda-xmlui/bitstream/handle/11250/2496177/Rapport%20nr%2086%20HVO%20Avis%c3%a5ret%202017.pdf?sequence=1&isAllowed=y, accessed June 12, 2019.

Hulme, Mike. 2017. *Weathered: Cultures of Climate*. London: Sage Publications.

Ingemark, Camilla Asplund. 2016. 'The Chronotope of the Legend in Astrid Lindgren's Sunnanäng'. In *Genre – Text – Interpretation. Multidisciplinary Perspectives on Folklore and Beyond*, edited by Koski, Kaarina, Frog and Ulla Savolainen, 232–250. Helsinki: Studia Fennica.

Ingold, Tim. 2007. 'Earth, Sky, Wind, and Weather'. *Journal of Royal Anthropological Institute* 13 (1) Issue Supplement: 19–38.

———. 2010. 'Footprint through the Weather-World: Walking, Breathing, Knowing'. *Journal of Royal Anthropological Institute* 16, Special issue: Making Knowledge: 121–139.

Joner, Selma, and Martin H. W. Zondag. 2018. 'Cicero-forsker: – Klimaendringer har gjort ekstremsommeren verre'. *Nrk.no*, July 19. https://www.nrk.no/norge/cicero-forsker_-_-klimaendringer-har-gjort-ekstremsommeren-verre-1.14131828, accessed January 8, 2020.

Kverndokk, Kyrre. 2019. 'Risk Perception through Exemplarity: Hurricanes as Climate Change Examples and Counterexamples in Norwegian News Media'. *Culture Unbound*, 11 (3–4): 306–329.

Kvistad, Yngve. 2018. 'Marerittet i sommerdrømmen'. *VG*, July 28, 2–3.

Mauren, Arnfinn. 2018. 'Flere skjønner alvoret nå'. *Aftenposten*, July 28, 4.

Mauren, Arnfinn and Wasim K. Riaz. 2019. 'SV: Regjeringen er handlingslammet i klimaarbeidet. Frp: Vi er heldige som har hatt en fin sommer'. *Aftenposten*, July 30, 17.

Nielsen, Bengt. 2018. 'Jeg ble fascinert og nesten litt redd'. *Nordlys*, July 20, 2018, 12.

Oring, Elliot. 2012. *Just Folklore: Analysis, Interpretation, Critique*. Los Angeles: Cantilever Press.

Solheim, Svale. 1952. *Norsk sætertradisjon*. Oslo: Aschehoug.

Søndeland, Geir. 2018. 'Statsministeren vil ha slutt på sommer-synsing'. *Stavanger Aftenblad*. June 27, 20.

Statistisk sentralbyrå. 2020. *Statistikkbanken*: Befolkning: 07459: Alders- og kjønnsfordeling i kommuner, fylker og hele landets befolkning (K) 1986–2020. SB. https://www.ssb.no/statbank/table/07459/tableViewLayout1/, accessed June 3, 2020.

Stephens, Michell. 2007. *A History of News*. San Diego: Harcourt Brace College Publishers.

Tvinnereim, Endre, and Kjersti Fløttum. 2015. 'Explaining Topic Prevalence in Answers to Open-Ended Survey Questions about Climate Change'. *Nature Climate Change* 5: 744–747.

Wurtele, M.G. 1971. 'Some thoughts on Weather Lore'. *Folklore* 82 (2): 293–303.

Yr. 2019. 'Været som var: Bergen'. https://www.yr.no/sted/Norge/Hordaland/Bergen/Bergen/statistikk.html, accessed July 2, 2019.

6 The prophetic tone in *True Detective*

Sensing the time of the future climate disaster

Isak Winkel Holm

Introduction

In 1984, in one of the darkest moments of the Cold War, French philosopher Jacques Derrida gave a talk at a seminar on 'nuclear criticism' at Cornell University, later published as the essay 'No Apocalypse, Not Now (Full Speed Ahead, Seven Missiles, Seven Missives)'. The *not now* of the title is to be understood not just as a supplication but also as a characterisation of the specific temporal structure of the Cold War: 'For the moment, today', Derrida writes,

> one may say that a non-localizable nuclear war has not occurred; it has existence only through what is said of it, only where it is talked about. Some might call it a fable, then, a pure invention: in the sense in which it is said that a myth, an image, a fiction, a utopia, a rhetorical figure, a fantasy, a phantasm, are inventions.
>
> (Derrida 1984, 23)

As Derrida argues, the nuclear apocalypse, even if it was *not now*, defined the experience of the actual *now* during the Cold War years: 'As no doubt we all know, no single instant, no atom of our life (of our relation to the world and to being) is not marked today, directly or indirectly, by that speed race' (1984, 20).

This chapter is concerned with the disaster which is *not now*. I write these sentences in a summer house outside of Copenhagen during the spring of 2020 in a moment where we, in the words of the Danish secretary of health, are standing 'at the foot' of an immense coronavirus outbreak. Even if the pandemic, for most people around the world, has existence only through what was said of it, and through what is shown on the omnipresent exponential graphs, it is marking every single instant, every atom of our life, for several months now. One day, when we are on the other side of the coronavirus outbreak, we will be standing at the foot of an even larger disaster which is also, in Derrida's apt phrasing, 'fabulously textual, through and through' (1984, 23). According to the 2018 report from the Intergovernmental Panel on Climate Change, global warming is likely to reach 1.5°C above pre-industrial levels between 2030 and 2052, causing not only sea-level rise but also desertification, droughts, hurricanes, wildfires, and epidemic diseases (IPCC 2018).

I will explore the temporal structure of the disaster which is *not now* by analysing the first season of *True Detective*, an HBO crime series authored by Nic Pizzolatto and directed by Cary Fukunaga that first aired in 2014. In the opening episode of the series, two Louisiana State Police Detectives Martin Hart and Rustin Cohle (played by Woody Harrelson and Matthew McConaughey) investigate the body of a female prostitute, Dora Lange. Kneeling in a prayer-like position in front of a lonely tree in a sugar cane field, naked except for a pair of deer antlers and a blindfold, she has, on her back, a spiral-shaped symbol about the size of a fist. In the subsequent episodes of the show, this tattoo-like symbol keeps showing up on bodies, buildings, and crime suspects. As it turns out, the symbol is the insignia of the mysterious sex cult that kidnaps and kills young women and children along the coast.

The investigation of the Dora Lange case takes place in a part of southern Louisiana devastated by violent hurricanes and monstrous oil extraction facilities. Detectives Hart and Cohle work on the murder case two times: first, in 1995, shortly after Hurricane Andrew, and then in 2012, not long after Hurricanes Katrina and Rita in 2005 and the Deepwater Horizon oil spill. In this context, the spiral symbol, shaped like a hurricane, is not only a clue about a past crime but also an omen of a future disaster. In other words, it is a sign that requires interpretation by a prophet as well as by a detective.

As it happens, Cohle is not only a very skilled detective but also a modernised and secularised version of a Hebrew prophet who predicts future disasters and provokes present authorities. In the second episode of the show, 'Seeing Things', he has a vision of a large flock of starlings that swirl into a spiral shape reminiscent of the symbol on Dora Lange's back. While we watch Cohle looking at the starlings, we hear him off-screen reflecting on his propensity to see things. Having developed a drug addiction during his assignment as a deep undercover narcotics agent before coming to southern Louisiana, he still suffers from chemical flashbacks and dreamlike hallucinations: 'then, the visions […]. Most of the time I was convinced that I'd lost it. But there were other times, I thought I was mainlining the secret truth of the universe' (E2:56).

Here Cohle emulates the prophetic calling that gave the prophets of the Hebrew Bible (or Old Testament) divine authority to reveal the secret truth of the universe. Isaiah had his lips cleaned with a live coal, Jeremiah was touched on the mouth by the Lord, Ezekiel was forced to eat a book scroll, and the twelve so-called minor prophets from Hosea to Malachi were initiated in similar ways. In Cohle's case, prophetic inspiration is prophetic injection.

At issue in this chapter, however, is neither *the prophet* as a social role nor *the prophecy* as a linguistic message but, rather, *the prophetic* as an aesthetic tone. Contemporary affect theory defines a 'tone' or an 'art mood' as a cultural object's 'affective bearing, orientation, or "set toward" its audience and world' (Ngai 2004, 43), or, in the same vein, as its 'affective character' (Plantinga

2014, 142–154). In similarity with, for instance, the ironic, the elegiac, and the apocalyptic tone, the prophetic tone can be described as an affective atmosphere with the capacity to determine not only a human subject's but also cultural object's set towards the world.

In what I suggest calling a 'prophetic tone', the perception of the precatastrophic present is affectively charged by the imagination of a postcatastrophic future. Together with the hurricane-shaped symbols on the crime scenes and the disaster-stricken landscape in the background, Cohle's prophetic utterings create the much-commented-upon 'ominous' mood of the show (Nussbaum 2014), that is, a certain way of seeing and sensing the world under the shadow of a future disaster. In Derrida's words, the catastrophic feeling-tone marks the series' relation to the world and to being. Thanks to this tonality, the first season of *True Detective* is not just a piece of well-made crime fiction but also, to my eye, a fascinating work of disaster fiction which is *naming* and *sensing* the time of a world threatened by virus outbreaks and climate disasters.

The core contention in the analysis that follows is that the concept of tone – a notoriously slippery one – should be located at the interstice between the sensible and the intelligible, between sensibility and rationality. Of course, an affective bearing is, first of all, a non-linguistic relation to the world and to being, but it is important to underline that it implies a conceptual interpretation of the world and of being as well: a tone is a matter of sensing *and* of making sense.

On the one hand, as we have seen, the sensible dimension of the prophetic tone is defined by the imagining of a future disaster. In this affective atmosphere, a human subject or a cultural object perceives the precatastrophic present from the vantage point of a postcatastrophic future. We recognise here a tonality premised on the imminence of disaster. On the other hand, as we shall see, the rational dimension of the prophetic tone is centred around the concept of justice in the broad sense of the term. When a human subject or a cultural object works backwards from an imagined future disaster, what comes into sight is most often conceptualised as a question of the normative order of the present world. Thus, as a hybrid of sensibility and rationality, of *aisthesis* and *logos*, the prophetic tone creates a particular nexus of disaster and justice.

In what follows, I first discuss the sensible dimension of the prophetic tone in the prophetic books of the Hebrew Bible and in *True Detective*, respectively. Thereafter, I turn to the rational understanding of justice, or of the ethical in the widest possible sense, as it is framed by this tonality.

The whole land quakes

Even if the sensible dimension of the prophetic tone is defined by the imagination of a catastrophic future, as I noted above, the tone is most often used to address a precatastrophic present. In the prophetic writings of the Hebrew

Bible, this complicated mix of future and present tends to be occulted by the common and erroneous understanding of the Hebrew prophet as a predicter, an understanding imported together with the Greek word *prophétes*. The Hebrew prophet, the *nabi*, does not only make statements about future disasters; he or she is also, and first of all, concerned with his or her contemporary society. If we focus on the prophetic as an aesthetic tone (rather than on the prophecy as a linguistic message), the disasters which abound in the prophetic books are not events that are predicted but, rather, images that play a role in the affective bearing of a text dealing with the present social and political situation. A similar argument could be made about the vast prophetic tradition inspired by the Hebrew Bible, but in this chapter I will restrict myself to the prophetic books.

Take Jeremiah as an example. At the beginning of the Book of Jeremiah, God calls the recalcitrant Jeremiah by touching his mouth and saying, 'Now I have put my words in your mouth' (Jer 1:9; I quote from NRSV). In the so-called Temple Sermon, a speech Jeremiah delivers standing 'in the gate of the Lord's house' (Jer 7:1), and for which, it seems, he is later arrested and tried (Jer 26:5), disaster is *not now*: Jerusalem is still thriving, Jeremiah can still take up his position in the gate of the Temple, but his prophetic initiation has made him able to hear the Babylonian army getting ready to invade Judah from the province of Dan in the North.

> We look for peace, but find no good,
> for a time of healing, but there is terror instead.
>
> The snorting of their horses is heard from Dan;
> at the sound of the neighing of their stallions
> the whole land quakes.
> They come and devour the land and all that fills it,
> the city and those who live in it.
> See, I am letting snakes loose among you,
> adders that cannot be charmed,
> and they shall bite you, says the Lord.
>
> My joy is gone, grief is upon me, my heart is sick.
> My grief is beyond healing.
> Hark, the cry of my poor people
> from far and wide in the land.
>
> (Jer 8:15–18).

Rather than depicting the future moment when the Babylonian army reaches Jerusalem and dreadful events unfold, Jeremiah's concern is the way in which 'the whole land' is presently trembling in anticipation of the disaster. In order to understand the sensible dimension of this tonality, it is helpful to break down the capacious concept of sensibility into two of its components, namely affect and imagination.

In the verses quoted here it is difficult to ignore the affective dimension – a potent mix of rage, terror, and above all grief: 'My joy is gone, grief is upon me, my heart is sick'. The mourning for the suffering of his fellow-countrymen is a characteristic feature of Jeremiah's prophesying, and, as in the passage quoted above, it can be difficult to determine whether the affect should be located in Jeremiah himself, in the Jewish people in general, or in YHWH. At times, the sorrow seems to be a structure of feeling in cultural scholar Raymond Williams's sense of the concept – that is, a collective affect expressed in the cries from the mourning rituals which can be heard far and wide in the land.

The imaginative dimension, on the other hand, creates a sensible image of the coming catastrophe. When Jeremiah hears the snorting and the neighing of the Babylonian horses, the auditive impressions are, evidently, imaginations rather than perceptions, given that the enemy's army is still camping in a province far from Jerusalem. Thanks to his prophetic gift, Jeremiah has the capacity to sense and feel the future disaster.

This imaginative dimension gives the prophetic tone its particular temporal structure. Jeremiah speaks about the precatastrophic present, but his attitude to the present situation is defined by his prediction of the future disaster. No single instant, no atom of Jewish life is not marked today, directly or indirectly, by the invasion to come. In grammatical terms, this structure could be described as future perfect (or future anterior) as opposed to the simple future tense. Used to describe an event that is expected to happen before a time of reference in the future, future perfect is a verb form that talks about the present moment as something that *will have been*. In the verses above, Jeremiah refers to Jerusalem as a city that will have been a city when the Babylonians come and devour it and those who live in it.

Interestingly, the affects mentioned in the Temple Sermon are conjugated in the future perfect. Regardless of how we choose to identify the 'me' who hosts the sorrow, it is remarkable that the sorrow itself is conjugated in the future perfect: it is a grief caused by an event which has not yet occurred. Some verses later, YHWH solicits his people to mourn and to raise a dirge over themselves: 'call for the mourning women to come' (Jer 9:17-10). Yet these skilled women will be tasked to mourn an event that lies in the future.

This temporal structure is pervasive in the prophetic books (and in the later prophetic traditions inspired by the Hebrew Bible). All the prophets predict the day of YHWH, the day of darkness and gloom (Joel 2:2), but their interest tends to rather lie in the *now* before the coming catastrophe; the moment when 'the day of the Lord is near' (Is 13:6). They lament for a life that will have been a life.

In modern philosophy, the temporal structure of the prophecy of doom has been explored by thinkers of nuclear and ecological disasters, most notably by Günther Anders, Hans Jonas, Jacques Derrida, and, most recently, Jean-Pierre Dupuy. According to Dupuy, the catastrophic future that awaits us demands that we take on an 'interpretative attitude' that perceives the present world in the light of future disaster (Dupuy 2002, 92). In Dupuy's view, the structure of this 'heuristics of fear' is a temporal loop (2005, 13): it 'invites us to make an imaginative leap, to place ourselves by an act of mental projection in the moment following a future catastrophe and then, looking back toward the present time, to see catastrophe as our fate' (2013, 33).

Jeremiah, for instance, makes an imaginative leap to the near future where he hears the Babylonian horses; then, from the vantage point of disaster, he looks back towards the present precatastrophic moment. Borrowing a concept from the French philosopher Peter Szendy, this temporal loop can be described as 'retroprospection' (2010, 29). Perhaps it would be more precise to talk about a proretrospection, since the loop starts out by an imagination of the future; in Gérard Genette's narratological terminology, the temporal loop can be described as an analepsis on a prolepsis (1980, 82).

Incidentally, the proretrospective structure of the prophetic tone can be explained by the complicated origin of the prophetic books. Since Jeremiah lived most of his life before the fall and destruction of Jerusalem, the bulk of the Book of Jeremiah describes the disaster as a future event. However, as contemporary philological research has documented, Jeremiah was not the sole author of the anthology that bears his name. In general, even if the prophets earned their authority as predicters of the future downfall of Israel and Judah, the prophetic books were transmitted and reworked by scribes in the centuries following the downfall of Jerusalem up until the second century BC. As biblical scholar Reinhard G. Kratz notes in a recent book, the narratives about Jeremiah take us to the precatastrophic times before the Babylonian invasion of Jerusalem and make an argument about the correct way to ensure the survival of the land and its inhabitants – but the Book of Jeremiah, as well as the prophetic tradition as a whole, 'is shaped by the past disaster – the fall of Israel and Judah – which the prophets interpret as God's judgment' (Kratz 2015, 26). The doom that the prophets predict lies already in the past (2015, 63). In other words, the temporal structure weaving together a precatastrophic and a postcatastrophic temporal perspective was created in a complicated writing process that took place both before and after the downfall. Whereas the prophets spoke prospectively about disaster, the scribes wrote retrospectively about it; taken together, the result was the retroprospective structure of the prophetic tone.

Under water within thirty years

Midway in the first episode of *True Detective*, Hart and Cohle leave the laboratory of the coroner after having examined the body of Dora Lange. Crossing the parking lot in a desolate suburb, Cohle makes a prophetic remark: 'This place is like somebody's memory of a town, and the memory is fading. It's like there was never anything here but jungle' (E1:29). In episode three, on one of the detectives' endless car rides through the destroyed landscape, Cohle, once again, takes the pose of a prophet of doom by predicting a future disaster: 'Pipeline covering up this coast like a jigsaw. Place is going to be under water within thirty years' (E3:44).

If the Hebrew prophets predicted future situations in which the two Jewish kingdoms were taken back by the desert, Cohle, their modern colleague, imagines a postcatastrophic moment in which the town is taken back by the jungle and the coast submerged by the bayou. In Jean-Pierre Dupuy's terms, Cohle makes an act of mental projection which places us in the moment following a future catastrophe. But then, from the imagined disastrous future, he looks back towards the present. Due to this temporal loop, Cohle is not looking towards the future but is, rather, oriented towards the present moment. The two prophetic remarks are explicitly about *this place*. In the first of them, Cohle even spells out that he perceives the present location as if he were an anonymous 'somebody' who, after some unnamed disaster, had only a fading memory of the town.

'Stop saying shit like that. It's unprofessional', Hart retorts when Cohle talks about the town being taken back by the jungle. Clearly, Hart is sensing that the tone of his colleague's remark is hopelessly out of tune with hard-headed police talk. Yet 'shit like that' plays an important role for the feel of the entire show. In other words, the catastrophic version of future perfect plays an important role not only for Cohle's prophetic utterings but also for the first season of *True Detective* as an aesthetic experience. To the viewer, the fictional southern Louisiana comes into view as a place that will be taken back by the jungle or will be under water.

This particular affective bearing is not only generated by the many verbal and visual references to past and future disasters but also by the narrative structure of the show. The first five episodes are organised by constant flashbacks and flashforwards between the original investigation in 1995 and the re-investigation in 2012. In the 1995 scenes, we meet Hart and Cohle as young police officers in great shape; in the 2012 scenes, however, we see them as worn-down and alcoholic ex-detectives and divorcees. The temporal loop of the frame story highlights an unusually brutal ageing process, thereby posing a question about the nameless disaster which, in the timespan between 1995 and 2012, transformed the two young professionals into middle-aged wrecks. Thus, thanks to the narrative structure of the show, the viewer experiences the young Hart and Cohle as men who are going to be suffering the effects of alcoholism and ageing bodies within seventeen years.

As we have seen, the historical model for this temporal structure is, first of all, the prophetic books of the Hebrew Bible, but Nic Pizzolatto also alludes to the nineteenth-century horror writer Ambrose Bierce. The name 'Carcosa', the sex cult's mythical headquarters, is borrowed from Bierce's short story 'An Inhabitant of Carcosa' (Bierce 1903). In this story, the narrator, the eponymous inhabitant of Carcosa, is attacked by a violent fever and wakes up in a bleak and desolate burial ground covered with a tall overgrowth of sere grass. Under the root of a tree he finds a decayed gravestone with his own name and the dates of his birth and death on it. As he soon realises, he has become a ghost lingering on in a post-apocalyptic future among the ruins of his ancient home city, another place taken back by the jungle. In Dupuy's words, he has made a mental projection to a moment following a future catastrophe.

We have sinned against the Lord

While the sensible dimension of the prophetic tone is defined by the image of a future disaster, the rational dimension hinges on the concept of justice. The Hebrew prophets *sense* the present moment under the shadow of the coming calamity, and they *make sense* of it as a breach of the covenant between the Jewish people and God. Indeed, the notion of a normative breakdown is pivotal to the interpretation of precatastrophic social and political life in the prophetic books. All the prophets are raging against a land that is full of bloody crimes and a city that is full of violence (Ez 7:23). Jeremiah's Temple Sermon can once again serve as an example. Before claiming that he is able to hear the snorting and the neighing of the horses of the Babylonian army in the far north, Jeremiah denounces Jerusalem as a city in which the order of justice is broken: '[T]here is nothing but oppression within her' (Jer 6:6). After this, he rises to a furious attack against those who believe that they are still adhering to the covenant:

How can you say, 'We are wise,
and the law of the Lord is with us,'
when, in fact, the false pen of the scribes
has made it into a lie?
The wise shall be put to shame,
they shall be dismayed and taken;
since they have rejected the word of the LORD,
what wisdom is in them?
Therefore I will give their wives to others
and their fields to conquerors,
because from the least to the greatest
everyone is greedy for unjust gain;
from prophet to priest
everyone deals falsely.
They have treated the wound of my people carelessly,
saying, 'Peace, peace,'
when there is no peace.
They acted shamefully, they committed abomination;
yet they were not at all ashamed,
they did not know how to blush [...]

Why do we sit still?
Gather together, let us go into the fortified cities
and perish there;
for the Lord our God has doomed us to perish,

> and has given us poisoned water to drink,
> because we have sinned against the Lord.
> We look for peace, but find no good,
> for a time of healing, but there is terror instead.
>
> The snorting of their horses is heard from Dan [...]
>
> (Jer 8: 8–16)

Here, Jeremiah diagnoses 'the wound of his people' as a normative breakdown: the law of the Lord is not with the Jewish people, it has been made into a lie by the pen of the scribes, the word of the Lord has been rejected by the people, and everyone is greedy for unjust gain. Interestingly, Jeremiah seems to be rather unconcerned with the content of the individual laws of Jewish society. He is not explaining how one should distinguish between lawful and unlawful acts or between just and unjust gain. Rather, he is focussing on the Jewish people's relation to the law in general – on the fact that law is losing its grip on the people.

This lack of concern for the content of individual laws is a recurrent feature in the prophetic books. The prophets seem to never tire of admonishing their listeners to do justice, *misphat*, by rescuing the oppressed, defending the orphan, and pleading for the widow, but they rarely reflect on the content of the laws that could be made in order protect the oppressed, the orphan, and the widow (Wolterstorff 2008, 67).

I want to suggest that the prophets' peculiar take on the concept of justice can be approached by a distinction, stemming from legal philosophy, between the *content* and the *force* of law. While the content of a law is a name for the specific acts that a given law prohibits or prescribes, the force of law denotes a law's capacity to apply itself to the world outside the legal system.

The prophet's disinterest in the content of the law has been discussed in the research since the sociologist Max Weber's seminal *Ancient Judaism* (1917–1919). The prophets were 'world-political demagogues and publicists', Weber writes, but he hastens to add that they were unconcerned with the subject matter of the political order: 'however, subjectively they were no political partisans. Primarily they pursued no political interests. Prophecy has never declared anything about a "best state"' (Weber 1967, 275).

In *The Prophets* (1962), the American professor of Jewish mysticism Abraham Joshua Heschel similarly distinguishes between the definition and the predicament of justice:

> The distinction of the prophets was in their remorseless unveiling of injustice and oppression, in their comprehension of social, political, and religious evils. They were not concerned with the definition, but with the predicament, of justice, with the fact that those called upon to apply it defied it.
>
> (Heschel 1962, 260)

The prophets were unconcerned with the definition of justice in the sense that they did not bother to discuss the content of the law. The Jews had entered into the covenant with YHWH in the Sinai desert, and, according to the prophets, the specifics of this God-given covenant were not negotiable. Instead, the prophets worried about the predicament of justice, as Heschel writes. A predicament is a difficult situation; some pages further down, for instance, Heschel writes about 'the predicament of man' as characterised by sin, guilt, and suffering (1962, 291). Hence, we can describe the situation in which justice finds itself in precatastrophic Judea as a predicament of justice. The prophets' strong sense of injustice is due to the fact that external circumstances make it impossible to apply the normative ideals of the covenant to concrete social life. In such a situation, God is imagined not as a guarantor of an order but as provider of a power that is able to wash away the obstacles to justice (Heschel 1962, 272).

More recently, political philosopher Michael Walzer has argued that the prophets, the first social critics in world history, chastise the rich for having forgotten 'not only the laws of the covenant but the bond itself, the principle of solidarity' (Walzer 1987, 85).

In 'The Force of Law: The "Mystical Foundation of Authority,"' an influential essay from 1990, Derrida discusses the force of law as a matter of its 'applicability' and 'enforceability' on the world (Derrida 1990, 924). Less well-known is legal philosopher Robert Cover's 1986 article 'Violence and the Word', in which he distinguishes legal interpretation from the interpretation of literature, from political philosophy, and from constitutional criticism. According to Cover, legal interpretation stands out by the coercive violence that ensures the 'embedding of an understanding of political text in institutional modes of action' (Cover 1986, 1601).

Thus, the concept of the force of law is situated in the gap that opens up between legal rules and the actual lives to which these rules are meant be applied – between law and its extra-legal other. According to the contemporary German philosopher Christoph Menke, who builds on Derrida and Cover, among others, the problem of law enforcement can be divided into two distinct problems: the problem is 'how to enforce the law not only against one who is unjust [*gegen den Ungerechten*], but also against one who is non-just [*gegen den Nicht-Gerechten*] – against one who stands outside, and is alien to, the justice of the law' (Menke 2018, 20). When law is enforced against one who is unjust – i.e. against the criminal person within the confines of the legal order – that enforcement deals with the well-known problem of illegality. But when law is enforced against one who is non-just – i.e. against a human being who is alien to the law, who turns his or her back on the law – it is responding to a zone of non-legality or extra-legality. In the perspective of the Hebrew prophets, their fellow countrymen were not unjust but rather non-just: their wicked acts were acts of apostasy and no mere criminal acts.

According to the prophet's grim political diagnosis, the law began to lose its force when the Jewish people reached the plentiful and fertile

Canaan and built the great cities of Samaria and Jerusalem. Betraying the historical moment in the desert when the people of Israel first entered into their covenant with God, the Jews began to worship Baal and the idols of neighbouring tribes, to sleep with Canaanite temple prostitutes, to sacrifice their children, and to make burnt offerings on the tops of the mountains. In this moral breakdown, the law became impotent and feeble, and the prophets' fellow citizens turned impudent, godless, hypocritical, and stubborn; they developed hearts of stone and turned their back, not their face, towards God (Ez 2:3; Ez 36:36; Jer 32:33). In Menke's terminology, this people is not unjust, but rather non-just – that is, inhabiting a zone outside the pale of law.

This is why the only solution to the predicament of justice is a 'turning back', a *teshuvah*, to a reinforced and revitalised version of the covenant. On rare occasions the prophets actually speak about a new law, but then their concern is not *what* should be written in the new law but rather *how* the new law should be written: in order to buttress the applicability and enforceability of the law, the new covenant should be written in the very heart of the Jewish people and not just on fragile stone tablets (Jer 31:33).

'They sacrifice kids and whatnot'

Like the Hebrew prophets, Detective Cohle is enraged by the breakdown of the order of justice in his present society. And like the prophets, this collapse shifts his attention from the content to the force of law. The inequities of the social order, in relation to factors such as class, race, and gender, seem to play no role at all for Cohle's understanding of justice. Instead, he focusses on the enforcement of justice. For years, women and children have disappeared in the state, but despite the enormity of the crime, the majority of police officers at the Vermilion Parish Sheriff's Department seem strangely reluctant to investigate it. After Dora Lange has been found dead in the field, in the first episode of the show, a group of police officers canvass door to door in the neighbourhood where she lived. As they return to the Sheriff's Department, Cohle smells alcohol on their breath and provokes them with a furious question: 'You guys canvass the bars pretty good today?' In Heschel's terms, Cohle is enraged by the predicament of justice in the state of Louisiana, by 'the fact that those called upon to apply it defied it' (Heschel 1962, 260). In line with the genre conventions of *noir* crime fiction, it is hinted that the police department is controlled by the Christian ministry magnate Billy Lee Tuttle, the presumed leader of the sex cult who also happens to be the cousin of Louisiana governor, Edwin Tuttle.

If we compare *True Detective* to the prophetic books of the Hebrew Bible, the infamous sex cult – a mixture of H. P. Lovecraft, voodoo, and Santerìa – is a modern US version of the Canaanite Baal cult. Like the Baal cult, the rituals of the sex cult rely on religious images, burnt offerings, and, above all, sacrifice: 'They sacrifice kids and whatnot', one witness reports (E4:3). A little later in the show, Cohle picks up on the idea of child sacrifice: 'You, these people, this place. It's like you eat your fuckin' young' (E6:22). In Menke's words, *these people* are not ones who are unjust but rather ones who are non-just; and *this place* is an extra-legal zone in which universal law has no pertinence.

Piece-of-shit-wise judgments

At the beginning of the third episode, Hart and Cohle have followed the trail of Dora Lange to a religious community, the Revival Ministry Church, where they attend a service in a tent. Like the camera, the conversation of the two detectives focusses on the people assembled in the tent rather than on the sermon held by the ecstatic pastor Joel Theriot. Whereas Hart, characteristically, regards the congregation with sympathy, Cohle, the pessimist, is suspicious of the institution of organised religion: 'I see a propensity for obesity. Poverty. A yen for fairy tales. Folks puttin' what few bucks they do have into a little wicker basket being passed around. I think it's safe to say nobody here's gonna be splitting the atom' (E3:5).

After a while, however, Hart and Cohle go on to discuss how to enforce the law in a group of people who have a yen for fairy tales. Christian belief plays an important role in this heated conversation, understood not as a religious faith but rather as a normative force, an institution that might possibly keep a person decent:

HART: Some folks enjoy community. A common good.

COHLE: Yeah, well if the common good's gotta make up fairy tales then it's not good for anybody. [...]

HART: I mean, can you imagine if people didn't believe, what things they'd get up to?

COHLE: Exact same thing they do now. Just out in the open.

HART: Bullshit. It'd be a fucking freak show of murder and debauchery and you know it.

COHLE: If the only thing keeping a person decent is the expectation of divine reward, then brother that person is a piece of shit; and I'd like to get as many of them out in the open as possible.

HART: Well, I guess your judgment is infallible, piece-of-shit-wise. You think that notebook is a stone tablet?

COHLE: What's it say about life, hmm? You gotta get together, tell yourself stories that violate every law of the universe just to get through the goddamn day. Nah. What's that say about your reality, Marty?

In Hart and Cohle's terse words, a 'piece of shit' is a human being who does not enjoy community and the common good but who, perhaps, might conform to the law thanks to the Christian promise of divine reward. Applying Menke's term, to be a piece of shit is to be one who is non-just: to be one who stands outside, and is alien to, the justice of the law. Accordingly, Cohle's judgment is not a legal or moral judgment that applies a specific law to a specific human being or a specific act. Rather, it is a judgment about the very applicability of laws. As we have seen, the Hebrew prophets passed judgments when denouncing their fellow citizens as 'pieces of shit' – that is, stubborn, stone-hearted, and godless – and as following the norms of the Canaanite Baal cult rather than obeying the laws of the covenant.

What Hart and Cohle disagree about is the nature of normative force supplied by Christian belief. The conciliatory Hart wishes people to stay decent. Without the *ersatz* normativity of Christianity, he argues, the community in the tent church would relapse into a lawless and normless chaos, 'a fucking freak show of murder and debauchery'. By contrast, Cohle, the radical, is unable to accept a decency which depends on made-up fairy tales. Rather than glossing over the chaos, he wants to get as many of the pieces of shit 'out in the open as possible'. The freak show of murder and debauchery is already in full swing, and there is no reason to hide it under a layer of superficial decency. This position is, of course, the position of the Hebrew prophet who insists on unveiling the hypocrisy of the existing order of justice. In Jeremiah's words, quoted above, what Cohle criticises Hart for doing is to say 'Peace, peace', when there is no peace (Jer 8:11).

The constitutional moment

'You think that notebook is a stone tablet?' Hart asks at the end of the discussion in the tent church, referring to Cohle's big black notebook which has played a conspicuous role in the earlier episodes. At the beginning of episode one, for instance, Hart explains that the officers at the Sheriff Department used to call Cohle 'The Taxman' because of this notebook: 'The rest of us had these little note pads or something. He had this big ledger. Looked funny walking door to door with it like the tax man, which ain't bad as far as nicknames go' (E1:7). In the tent church, when Hart refers to the notebook as a stone tablet, he ironically turns Cohle the taxman into Cohle the lawgiver, a modern-day Moses descending from Mount Sinai, carrying the stone tablets of the law. Whereas a taxman is maintaining a given social order, a lawgiver is creating a new social order by turning a crowd of pre-political human beings into a political community. In Jean-Jacques Rousseau's words, Moses made the astounding deed of transforming a herd of wretched fugitives into 'a body politic, a free people' (Rousseau 1972, 6).

The constitution of the Jewish people at the foot of Mount Sinai is a frequent theme in the prophetic books of the Hebrew Bible. For instance, in The Book of Hosea, the Lord says: 'I am the LORD your God / from the land of Egypt; / I will again make you dwell in tents, / as in the days of the appointed feast' (Hos 12:9). In other words, there is an interesting connection between, on the one hand, the prophets' looking forward towards the catastrophic moment in the future and, on the other hand, the prophets' looking backwards towards the constitutional moment in the past. The *teshuvah* of the prophets, their wished-for 'turning back', is also a return from the city to the desert.

If we interpret Hart's sarcastic remark in the light of the prophetic books, the sarcasm might also hide an important insight into the normative breakdown that plays a role in *True Detective*. If the fading laws of southern Louisiana should end up by losing their grip on the citizens of the land, if people should finally end up as pieces of shit 'out in the open', it would surely be 'a fucking freak show of murder and debauchery'. However, this moment of non-legality could also, potentially, turn into a constitutional moment in which a new covenant is instituted. According to Derrida, the performative force of law was present in 'the very emergence of justice and law, the founding and justifying moment that institutes law' (1990, 941).

This conflict perhaps explains the choice of location for Carcosa, the mythical headquarters of the sex cult and the place of the showdown with Dora Lange's murderer, Errol Childress, in the last episode of the show. On the real map of Louisiana, the scenes at Carcosa are shot in the ruins of Fort Macomb, a stronghold built in 1822 as a defence against the British invasion of Louisiana and later reused by both Confederate and Union forces during the Civil War. In other words, the historical Fort Macomb had a part to play in the two conflicts that, more than any other, defined the United States. Thus, returning to Fort Macomb is the modern-day Louisiana version of returning to the desert where the Jewish people entered into the covenant. In this case, however, the location of the site of the foundation is not a clean desert but an untidy ruin. The corridors of the fort have become stuffed with vegetation, mysterious symbols, and embalmed bodies of murder victims. It takes the extra-legal force of Hart and Cohle, at this late moment acting as ex-police, to clean up the site of the political constitution.

The black stars rise

In episode five, Detectives Hart and Cohle find Reggie Ledoux, who they falsely identify as the murderer of Dora Lange, in a remote farm in the marshlands and execute him while he is still handcuffed, without due process of law. In the last moments of his life, Ledoux makes a number of ominous apocalyptic remarks: 'It's time, isn't it? The black star', 'The black stars rise', 'I know what happens next. I saw you in my dream. You're in Carcosa now with me', and 'Time is a flat circle' (E5:13). The black stars allude to the Revelation to John where, as the fourth angel blew his trumpet, 'a third of the sun was struck, and a third of the moon, and a third of the stars, so that a third of their light was darkened' (Rev 8:12).

'What's that, Nietzsche? Shut the fuck up', Cohle replies, in a sense repeating Hart's rejection of his own prophetic sayings in the first episode ('Stop saying shit like that. It's unprofessional'). But here, too, the tonality of the utterings lives on in the show. When the Dora Lange case is inspected by two higher-ranking detectives in 2012, Cohle repeats Ledoux's strange mixture of apocalypticism and Nietzscheanism: 'This is a world where nothing is solved. Someone once told me, "Time is a flat circle." Everything we've ever done or will do, we're gonna do over and over and over again' (E5:19).

One way to put the claim I am making in this chapter is that we need to distinguish between prophetism and apocalypticism. As we have seen, both the prophetic and the apocalyptic tones play a role in *True Detective*, but my contention is that there is a crucial difference between these two distinct affective bearings. In fact, one of the reasons that the prophetic tone is an understudied category is that it tends to be conflated with what is sometimes called an apocalyptic tone (Derrida 1982) or an oracular tone (Nietzsche 1988, 369). Let me sum up the above analysis of the prophetic tone in *True Detective* by contrasting it with the apocalyptic tone.

Even if St John is often referred to as a prophet, prophetism and apocalypticism belong to two different epochs of biblical history. Jewish apocalypticism emerged in the books of Enoch and Daniel in the Hellenistic period, whereas the most famous contribution to Christian apocalypticism, the Book of Revelation, was written in the first century AD, more than 500 years later than, for instance, the historical Ezekiel, the youngest of the three major prophets.

More importantly, the prophetic tone offers an assemblage of sensibility and rationality that differs from that of the apocalyptic tone. At the level of sensibility, both tonalities can be described as temporal structures, yet it is easy to point out the dissimilarity of the two structures. As we have seen, the sensible structure of the prophetic tone is a temporal loop, the mix of prospection and retrospection which I have analysed above. This tonality is intrahistoric and immanent in so far as it weaves together two moments in time, a precatastrophic present and postcatastrophic future (Ricoeur 1994, 84; LaCocque and Ricoeur 1998, 170). By contrast, the apocalyptic sensibility is not *a loop in time* but rather a *leap out of time*. The Greek word *apokalúptō* literally means to unveil that which is hidden: an atemporal and ahistorical layer of reality which is disclosed only in the 'end times' when human history has reached its conclusion. In their apocalyptic mode, Ledoux and Cohle sense time as a flat circle.

At the level of rationality, on the other hand, both the prophetic and the apocalyptic tones trigger a discussion of the concept of justice. Both prophetism and apocalypticism are centred around normative breakdowns. However, we must recognise here two normative breakdowns and, hence, two dissimilar notions of justice. As discussed above, the prophetic tone raises a question about the force of law. In this context, the law is to be understood as the covenant entered into in the Sinai desert. Even if the prophets do not discuss the content of the individual moral and legal rules that make up the covenant, their concern is the existing order of justice in their contemporary society. The apocalyptic tone, on the other hand, implies no awareness of human justice. In the cosmic drama of the Last Judgment, the breakdown of social norms leads to a shift in focus from human justice to divine justice, and seen from this metaphysical perspective, any particular historical covenant, and its force or its weakness, is obsolete.

The point I am making is that the distinction between a prophetic and an apocalyptic tone is crucial not only for the exploration of works of fiction such as *True Detective* but also for an understanding of the temporality of the coronavirus outbreak and the climate catastrophe. The two tones should be approached as two cultural conditions of possibility for the way in which we are sensing time in a world threatened by disaster which is *not now*. In the prophetic tone, we perceive time as an intrahistorical interval of urgency in which ethical human action is possible and called for. If we frame the future disasters as apocalyptic, on the other hand, the moment of action disappears in the flat circle of mythical time.

References

Bierce, Ambrose. 1903. *Can Such Things Be?* Washington, DC: The Neale Publishing Company.

Cover, Robert. 1986. 'Violence and the Word'. *Yale Law Journal* 95 (8): 1601–1629.

Derrida, Jacques. 1982. 'Of an Apocalyptic Tone Recently Adopted in Philosophy'. *Semeia* 23: 63–97.

———. 1984. 'No Apocalypse, Not Now (Full Speed Ahead, Seven Missiles, Seven Missives)'. *Diacritics* 14 (2): 20–31.

———. 1990. 'Force of Law: The "Mystical Foundation of Authority"'. *Cardozo Law Review* 11: 919–1045.

Dupuy, Jean-Pierre. 2002. *Pour un catastrophisme éclairé: quand l'impossible est certain.* Paris: Seuil.

———. 2005. *Petite métaphysique des tsunamis.* Paris: Seuil.

———. 2013. *The Mark of the Sacred.* Trans. M. B. DeBevoise. Stanford, CA: Stanford University Press.

Genette, Gérard. 1980. *Narrative Discourse: An Essay in Method.* Trans. Jane E. Lewin. Ithaca, NY: Cornell University Press.

Heschel, Abraham Joshua. 1962. *The Prophets.* New York: Harper & Row.

IPCC. 2018. *Global Warming of 1.5°C.* https://www.ipcc.ch/sr15/

Kratz, Reinhard G. 2015. *The Prophets of Israel.* Trans. Anselm C. Hagedorn and Nathan MacDonald. Winona Lake, Indiana: Eisenbrauns.

LaCocque, André, and Paul Ricoeur. 1998. *Thinking Biblically: Exegetical and Hermeneutical Studies.* Chicago, IL: University of Chicago Press.

Menke, Christoph. 2018. *Law and Violence: Christoph Menke in Dialogue.* Trans. Alessandro Ferrara. Manchester: Manchester University Press.

Ngai, Sianne. 2004. *Ugly Feelings.* Cambridge, MA: Harvard University Press.

Nietzsche, Friedrich. 1988. *Nachgelassene Fragmente 1869–1874.* Ed. Giorgio Colli and Mazzino Montinari. Vol. 7, *Sämtliche Werke: kritische Studienausgabe in 15 Bänden.* München; Berlin; New York: Deutscher Taschenbuch Verlag; De Gruyter.

Nussbaum, Emily. 2014. 'Cool Story, Bro'. *The New Yorker,* March 3, 2014.

Plantinga, Carl. 2014. 'Mood and Ethics in Narrative Film'. In *Cognitive Media Theory,* edited by Ted Nannicelli and Paul Taberham. New York; London: Routledge, Taylor and Francis Group.

Ricoeur, Paul. 1994. 'Philosophie et prophétisme 1.' In *Lectures 3: aux frontières de la philosophie*. Paris: Editions du Seuil.

Rousseau, Jean-Jacques. 1972. *The Government of Poland*. Trans. Willmoore Kendall. Indianapolis: Bobbs-Merrill.

Szendy, Peter. 2010. *Prophecies of Leviathan: Reading Past Melville*. New York: Fordham University Press.

Walzer, Michael. 1987. *Interpretation and Social Criticism*. Cambridge, MA: Harvard University Press.

Weber, Max. 1967. *Ancient Judaism*. Trans. Hans Gerth and Don Albert Martindale. New York: Free Press.

Wolterstorff, Nicholas. 2008. *Justice: Rights and Wrongs*. Princeton, NJ: Princeton University Press.

7 Advocating equilibrium

On climate change at public aquariums

Lars Kaijser

Introduction

Placed on a dock along the Tagus River, the grey concrete building hosting Oceanário de Lisboa rises five storeys, at least, in height. When entering the aquarium, visitors have to use a footbridge, connecting the large building with a house on land that lodges the entrance, restaurants, and a shop. Along the footbridge, displays detail the conservation work supported by the aquarium. These promote projects aiming to reduce human impact on the environment, such as declining seagrass beds and seahorse populations, and problems of overfishing, littering, and bycatch. Added to this is information on how the aquarium raises awareness of the present amphibian situation. Other displays explain how the oceans cover 70 percent of the earth's surface and contain 90 percent of the earth's biomass. The final display also states the twofold mission of the aquarium: 'To promote the knowledge of the oceans and raise awareness about our duty to conserve nature by changing our behaviour'.

The Oceanário de Lisboa opened in 1998 as part of the 1998 Lisbon World Exposition.[1] Upon its opening, the aquarium was the largest in Europe and was hailed as an example of the new kind of spectacular aquariums established during the last decades of the twentieth century, or, as stated in a book on the history of the public aquarium, the Oceanário de Lisboa represents 'the extreme, radicalised form of a good old aquarium' (Brunner 2011, 133). The aquarium in Lisbon has been successful and has on three occasions been named the best aquarium in the world by TripAdvisor: in 2015, 2017, and 2018.[2] The Oceanário de Lisboa had 1.3 million visitors during 2018, engaging 168,000 children and adults in educational activities. Like several similar public aquariums, the aquarium in Lisbon is a nexus for disseminating knowledge of nature and conservation. Defined as part of the wider genre of popular science, the public aquarium presents simplified scientific insights for a broader audience (cf. Chapter 5 and 8. With a focus on climate change, this chapter will take a closer look at the way that the human impact on marine environments is staged at the public aquarium.

Stories and temporalities

There are several reasons for choosing the Oceanário de Lisboa for this chapter. Being a target for school classes and a highly valued tourist attraction with a large number of visitors justifies a study, but there are other reasons. My visit to the aquarium came at the end of a research project on how public aquariums stage information on nature and environmental issues.[3] What I encountered at Oceanário de Lisboa came as no surprise; on the contrary, since the exhibits displayed more or less the same information as other facilities I had visited, it was a good example of how aquariums stage environmental challenges.[4] To begin with, it selected expected animals and habitats for display, and it represented these typical choices in line with contemporary ways of disseminating information and viewpoints through storytelling (Gröppel-Wegener and Kidd 2019, 8). Here, story is understood as a way of organising maritime facts together into a more coherent presentation carrying a meaningful statement, making sense of different accounts of the marine world. This is also how the curation of exhibits has been presented to me during interviews and discussions when visiting different aquariums.[5] The layout of the Oceanário de Lisboa – in content and narrative organisation – was also an example, as in exemplary, of the way that environmental stories tend to be staged, summarising my experiences of exhibits at other aquariums. There is a striking clarity and simplification in the figuration of nature, environment, and conservation in this facility. Its exemplary quality makes it a good starting point for a more detailed discussion of storytelling, environmental threats, and popular science.

To investigate the stories told at the aquarium, this chapter will apply a set of concepts taken from narrative researchers William Labov and Joshua Waletszky's model for studying oral narratives of personal experiences (Labov and Waletzky 1967). This theoretical model helps to establish how the different exhibits at the aquarium play a part in conveying a message, and to identify what this message is. The inspiration for using Labov and Waletzky stems from discussions with aquarium staff and from my observation that similar patterns of presenting environmental messages tend to reappear at different aquariums. The stories told at the aquarium are not personal experiences; instead, they represent scientific insights told to inspire and to cause action. Due to the nature of this content, this chapter will scrutinise how the aquarium stages environmental changes rather than how visitors experience the same story. It does not offer an investigation of the curators' intentions; it is firmly directed towards an interpretation of the environmental stories identified on my walk through the aquarium that importantly – because these stories and the way they are presented recalls other aquariums – represent a common approach to staging environmental changes. The spatial organisation at Oceanário de Lisboa makes it a good test case for how Labov and Waletzky's model can be adapted for the analysis of exhibits.

According to Labov and Waletzky, a sequence of events makes a story: a story describes a sequence in time, giving structure and meaning to a depicted occasion, and the structure of the story influences the point that the story tries to make. The model consists of six concepts. The *abstract* introduces a narrative, explaining the reason for telling the story and identifying the main point. The *orientation* provides information on time, place, persons, activities, and situations. *Complicating action* helps to create tension, generating an interest in the story. *Evaluation* emphasises the point of the story and what hopefully will engage the audience. The *resolution* ends the story, identifying how things turned out. The *coda* arrives after the end of the story, taking the listeners back from the world of the story, and into the now and the storytelling situation. In the following discussion, this model helps to conceptualise the spatial organisation of the aquarium, and how displays, habitats, and other props appearing along the path in different ways contribute to an environmental message. This understanding of how a story is told relies on the movement between equilibrium and disequilibrium, and claims that the purpose of the story is to restore equilibrium (Labov and Waletzky 1967; cf. Palmenfelt 2017, 38–45).

Climate change per se is not the subject matter of the Oceanário de Lisboa's story, but the conditions of marine environments is, and climate change is part of those complex conditions. At the core of this story is the aquarium's aim to change people's behaviour. In this setting, the recreation of place is an important matter. This is, indeed, the general idea of an aquarium, which typically recreates marine localities displaying the characteristics of habitat with significant geological formations, animals, and plants. However, aquariums also tell stories that emphasise temporality: often, this could be deep-time stories of how long different species inhabited the earth or stories involving rapid contemporary environmental change. Such displays, then, reveal how the coordination of recreated places together with different temporal frameworks might affect the understanding of climate change. It is this association of time and place that makes up what Mikhail Bakhtin has identified as a chronotope (Bakhtin 1991, 14). As presented in the introduction to this book, Bakhtin's chronotope conceptualises the amalgamation of time and space as a foundation through which the rest of the story and its various characteristics can develop (Bakhtin 1991). Stories at the Oceanário de Lisboa are constructed through combinations of signs, animals, staged environments, photos, and short films – a range of media that suggest two different chronotopes: the specific habitat with its cyclical time, and global climate change (tied to the pace at which the global environment alters). This is also the fundamental argument for this article, where the theoretical approaches from Labov and Waletzky and Bakhtin complement each other. The narrative model helps to make sense of the narration of environmental change, its display and its resulting message, while the concept of the chronotope serves to identify how differences in the alliance of place and temporality have a significant impact on this message. The chapter will continue with a walk through the oceanarium, tracing climate change through various exhibitions.

Walking the story

Organised through a 'one ocean-concept', the main building at the Oceanário de Lisboa is designed with a large cubic tank placed in the centre, and four staged habitats in each corner. The tank rises from the lower floor and seven metres up to the ceiling of the upper level, connecting all public parts of the building.[6] The public space comprises two floors arranged to form a one-way route, with a path leading counterclockwise around the tank from the entrance on the upper floor to the exit on the lower. On the upper floor, the path moves along the staged shoreline of the four habitats while on the lower level, the path continues as an imaginative walk beneath the water surface, this time with the habitats viewed from below.

The story at an aquarium unfolds when walking through the premises, joining the different dots of displays with texts, pictures, statistical diagrams, and maps, together with living animals and staged habitats. It is a story experienced through movement and encounters with different environments and messages. When following the devised path, the large ocean tank is always on the visitor's left side. Imbued with the same spirit characterising most contemporary aquariums, natural history museums, and similar facilities, a walk around the large tank is made to evoke both curiosity and fascination for nature and its challenges (Kaijser 2019). This effect is primarily due to the organisation of the aquarium space. The ocean tank can be viewed from different angles, sometimes through large windows two storeys in height, sometimes through smaller alcoves allowing only one or maybe two visitors to glance at the animals inside. The same goes for the recreated habitats, seen from both above and below. The path meanders through different habitats of the marine world, tied together through slightly narrow and sometimes darker passages.

A story is often defined as having a beginning, a middle, and an end. This applies to the aquarium as a whole, with an entrance at the beginning and an exit at the end efficiently bracketing the story, as well as to individual habitats or exhibits. As described at the beginning of this chapter, the aquarium offers a kind of abstract of its story in the footbridge that visitors use to access the rest of the building. The footbridge leading into the aquarium contains the institution's general message, which is that aquatic environments are threatened by human activities and humans need to change their behaviour to save the world. The mission statement works as a *key text* framing the overall story, hinting at possible interpretations (Ingemark 2016). The different displays discussing conservation work contain a global approach that describes the earth and different ecosystems in a general way. One of the signs discusses amphibians, showing a picture of a green frog, stating how amphibians play a key role in the balance of the ecosystem, and how the aquarium works to raise awareness about the threats to amphibians through an exhibit and financial support to conservation projects. Importantly, the presentation of the amphibian work localises the environmental changes in question to Portugal and the Iberian Peninsula.

To sum up, to walk up the footbridge is to walk through the aquarium's abstract and into the stories it has to tell (likewise, when at the end, leaving the aquarium building on the same footbridge is to walk out of the story). The signs clarify that the marine environment covers most of the earth, that the earth is threatened by environmental changes caused by humans, that these threats are here and now, and that the aquarium works in different ways to restore the balance of the ecosystems.

Orientations within cyclical time

The first room in the aquarium building is dark, with a large window opening into the large tank. In front of the slightly crescent-shaped window, visitors gather, and behind them sharks, rays, a sunfish, barracudas, groupers, and shoals of other fishes circulate in the bluish water. A sunbeam – artificial or real, I do not know – shines through the water, reflecting on the aquatic creatures swimming through. At the side of the large glass window is a map displaying the route around the aquarium and its four habitats, namely the North Atlantic, the Antarctic, the Temperate Pacific, and the Tropical Indian. The map gives a good overview of the geographical areas covered, but – as this chapter will show – it did not include the environmental exhibits.

As stated, the mission of the aquarium is to promote knowledge of the marine environment and to raise awareness of the need to change behaviour. There are some important differences in the way these two missions are handled. The promotion of knowledge comes in facts and figures, as well as in short story-like representations of relationships within nature. To a degree, these representations consist of enumerations (i.e. specific characteristics of an animal), which is an account of the world but not a story (cf. White 1980). This is different from the way that the need for changed behaviour is presented, where the use of storytelling is pivotal.

The importance of changed behaviour is tied to an overall story told at the aquarium, what I will call the *general environmental story*. The general story of the aquarium evolves along the path through the aquarium, informing visitors of environmental threats to a world in balance. This is not the only story presented at the aquarium; there were others. For example, stories of the history of the Oceanário de Lisboa, or stories characterising individual animals or habitats. As in many narrations, the general environmental story consists of several simple narratives – with both orientations and complications (cf. Labov and Waletzky 1967, 32). As mentioned, the orientations in a story help to identify persons, places, situations, and activities important to the telling. Here, information, animals, and artificial scenery help to build up an understanding of the individual habitat. Important orientations are information on animals, nature, and annual cycles, imbuing the visitor with a sense of how these habitats feature forms of balance. The main character in the general environmental story, however, is not the animals, but the threatened habitats and the importance of balance within this context.

As said, the story evolves when moving through the aquarium, reading signs, looking at animals, immersed in aquatic environments. After leaving the entrance, through a short, dark passage, the display of the North Atlantic and the first habitat opens up. The public area resembles the inside of a cave or a crevasse with openings towards a landscape with water, rocks, trees with green leaves, and a small waterfall pouring from a cliff edge. Atlantic puffins sit on the rocks and a common murre is swimming and occasionally diving into the depths of the water, with sounds from the waterfall and bird calls contributing to the experience. On the opposite wall towards the staged habitat are information displays, presenting the habitat as representing the North Atlantic coastal area, from the cold northern reaches of the British Isles to the volcanic Azores islands. A pattern for the displays at the aquarium is to feature two paragraphs, with the first giving an orientation into the habitat or to the animals and the second presenting environmental threats. The display lists the birds in the habitat with photos and an introductory text in Portuguese and English providing some basic facts about the birds. The text begins by noting the Alcidae family before naming three species of birds of the habitat that belong to it: Atlantic puffin, razorbill, and common murre. Added to this is a distribution map showing their location in the wild, and their conservation status, which in these cases is 'least concern'. Another display gives information on the birds' breeding patterns, and how they can travel up to 300 kilometres searching for food. The display also states that 'Today tons of plastic trash swirls on the ocean currents. Seabirds looking for flashing fishes frequently mistake shiny debris for food. With their stomachs full of plastic instead of fish, many oceanic birds risk starvation'. The displays in the Atlantic area are representative of most of the displays at the aquarium. They give a general overview of the habitat and present the animals according to eating habits, breeding and migration patterns, and areas of habitation. There is also a tendency to point out some of the more astonishing and entertaining facts of the animals – the so-called wow factor – such as a bird's capacity to fly 300 kilometres for food.

The typical way for aquariums to organise knowledge of marine environments is as above, through the chronotope of the habitat, with specific flora and fauna as setting, characterised by both biological and physical factors determining the parameters for life, and a specific temporality. The staged habitats are idealised versions of nature: not actual places but spaces of nature reduced to its basic features. Favoured habitats tend to focus on striking natural features such as the coral reefs or rainforests, or charismatic animals such as otters, sharks, or penguins. The habitats at Oceanário de Lisboa answer to the general criteria of representing habitats but with a selection of preferred natural environments and a local connection. The Antarctic and the tropical habitat with a rainforest and coral reefs represent charismatic nature, while the North Pacific displayed a charismatic animal, the sea otter. Most aquariums also represent local flora and fauna with habitats showing regional or national ecosystems; at least, this is my impression after having visited several aquariums. At the Oceanário de Lisboa, the local waters were represented by the North Atlantic, covering the Azores – an archipelago of volcanic islands, constituting an autonomous region of Portugal.

The chronotope of the habitats also distinguish themselves through their temporality (cf. Chapter 9). They manifest a cyclical time in the presentation of the animals' behaviour, staging the circle of life when the animals eat and breed. The habitats in themselves are more or less pleasant-looking environments without any human impact, tending to recreate a natural aesthetic recognisable from travel advertisements or similarly picturesque scenery. Occasionally deep-time references appear stating a continuity, such as coral reefs being around since 500 million years ago. This primary understanding of nature recognises it as something for which temporality comes in the cycles of night and day, summer and winter – something built on periodic repetition – and as something that has been present for a very long time. It is this image of balanced nature, with a repetitive, cyclical temporality, that the aquarium strives to protect.

The staged habitats work as narrative orientations within the general environmental story, presenting natural environments and animals with certain qualities worth preserving and following a pedagogical ideal built on immersion (cf. Karydis 2011; Kaijser 2019). The visitors should feel the exhibits and a sense of being part of a staged habitat, ideally resulting in interest and empathy for the aquatic environments. The idea of the habitat is rooted in contemporary science and based on the notion that the best way to understand plants and animals is in their context (Doordan 1992). Also, aquariums tend to favour environments that are free from human interference. Even though it is possible to find aquariums diverting from this in certain details – not featuring all of the animals or props expected in a represented habitat, for example – the overall pattern seems to reappear. The knowledge imparted in a given habitat has a value of its own, but in relation to the general environmental story told at the aquarium, it has a distinct purpose: it displays a nature worth preserving. The chronotope of the habitat is matched with another chronotope, one of environmental change, which will be presented in the next section.

Complications in the dark

As described above, the map at the entrance of the aquarium shows the layout of the aquarium, with the main tank in the middle, and four habitats located in each corner. Left out of this overview were four exhibits detailing environmental issues. Thus, the map relates the staged habitats to one another and presents a route through the aquarium; however, it does not cover the environmental story in a similar way. To apply Labov and Waletzky's model of how stories are organised, the map features orientations but excludes complications. According to this narrative model, complications ignite the story with tension, often making the story worth telling by introducing a disequilibrium that needs to be resolved. Like climate change disturbing the balance of nature, the complication instigates imbalance into the aquarium's general environmental story.

The narrative complication at the aquarium consists of threats to the balance in nature, and this information appears throughout the aquarium. At first, it is possible to detect this message on the displays in the habitat areas. Identifiable as a complicating action, the second paragraph on the displays create tension, unbalancing the staged habitat and alerting visitors to the conservation message. This was noted earlier in the quoted example from the North Atlantic habitat discussing how 'Today tons of plastic' threatens the ocean environment. Though the problem of marine plastics receives less attention in this article, at the public aquarium it becomes a pivotal environmental challenge.[7] When it comes to climate change, most staged environments could inform the public on this issue; however, the aquarium tends to tie climate change to the acidification of the ocean, deforestation in the rainforests, and ice melting in the Arctic and the Antarctic. This was the case at the Oceanário de Lisboa.

The first environmental exhibit, called 'One Planet, One Ocean', tells of ocean acidification, climate change, and the human footprint on the ocean. The exhibit opens with a voice reading from Carl Sagan's *Pale Blue Dot: A Vision of the Human Future in Space*, describing the earth as a tiny fragment in a vast universe, but also the only place to live that we know of (cf. Sagan 1994). Starting with this interstellar chronotope, the One Ocean exhibit consists of tables set with, among other things, shark specimens such as jaws, teeth, and skin laid out for the visitors to inspect up close, and touchscreens informing on ocean temperature, colours, currents, tsunamis, and the global conveyor belt. The screens show global maps of the world, and human impact on the oceans is traced through changing colours, using a generic scheme with green or blue as good and red for danger.

Scale and numbers underlined the arguments, describing the connection between environmental changes, life in the ocean, and sustainable living. Important to note is the stress placed on balance, conceptualised on displays in the exhibit and through the use of words such as 'imbalance' or clauses such as 'disrupt the delicate balance of species'. In relation to the habitats, there is a difference in time signature: stability and continuity is not the dominant mode, but rather rapid change. If the habitats stage isolated parts of the world, the environmental exhibits present a global perspective, which produces a significantly different chronotope. It fosters a chronotope with a worldwide approach, where global climate change connects with local circumstances. The presentation of environmental crises traces these to human activities, explaining how the planet has entered climate change and how the understanding of this is based on scientific procedures and knowledge. In this way, the environmental exhibits also work as an intertext – as a network of associations – providing information that helps visitors to understand and to connect with that information and that hints at the environmental concerns found on displays in the habitat areas (cf. Ingemark 2016).

The patterns found in the One Ocean exhibit continue in the other environmental exhibits. On the lower level, in a dark corner, is an exhibit on amphibians. During the last decades, the amphibians' situation has gained much attention, with numerous amphibian species on the brink of extinction. There are several threats to their survival: changes in water and soil quality, loss of habitat, climate change, and the chytrid fungus, which has spread rapidly among the amphibians. Regarded as an indicator for the well-being of the environment, as they depend on both land and water habitats, amphibians have gained increased attention from public aquariums, and during my project, I have attended exhibitions dedicated to alerting the public to the amphibian situation in Vancouver, Long Beach, Chicago, and Stockholm. The amphibian class of species has become a key class for discussions of environmental changes and the human impact on nature, and for relating global issues to local conditions.

The engagement with amphibians is central to how the Lisbon aquarium presents itself and, as mentioned before, it appears early on in the 'abstract' offered by the footbridge leading to the aquarium building. Though not on the map, it is a permanent exhibit, located in a murky, cramped area with terrariums, props, and signs, divided by screens showing pictures of frogs, toads, salamanders, and newts. It is difficult to get a sense of the layout of this space as it is under construction, and when trying to move around the premises, I occasionally bump into other visitors. The exhibit gives a general overview of the global amphibian situation, while at the same time localising the issue to Portugal. In many ways, this articulates a stance on environmental issues. For instance, there is no doubt here of who is responsible for climate change, or that climate change is taking place. In this way, the aquarium has taken the same stand as most scientists.

The exhibit starts with a black screen with a close-up photograph of a green frog hiding behind a leaf, the red eyes peeking above the edge, and a text saying 'Amphibians. Interesting by nature'. The exhibition on amphibians is divided into general presentations of different habitats – *Still Water, Forest Ground, Moving Waters* – demonstrating the wide variety of habitats that amphibians inhabit, and their ability to adapt to different conditions. There is also a display presenting the local amphibians, the Iberic habitat, with a unique set of amphibian species living nowhere else in the world: the golden-striped salamander, the Iberian frog, and the Iberian midwife toad. These species are not on display, but other amphibians from other parts of the world are, such as South American poison dart frogs and red-eyed tree frogs. The conservation projects engaging the Oceanário de Lisboa are accounted for with presentations of a micro-reserve for amphibians in the region of Mindelo, constructions of small ponds, and educational activities. The information is both general and sparse in detail, but it helps to resituate the understanding of the environmental crisis from the general to the local and specific, from a global matter to a neighbourhood issue. A picture with a green frog similar to the image at the beginning also waves goodbye to visitors with the following statement: 'The global loss of amphibians reflects how our planet is changing. It is also a warning that only by changing our behaviours can we ensure the sustainability of the planet'.

The amphibian exhibit combines the two major chronotopes found at the Oceanário de Lisboa. Bakhtin has defined the coexistence of different temporalities as *multi-temporality* or *synchrony*; this is also what characterises the chronotope at the public aquarium, with different temporalities present at the same time. The amphibian exhibit frames the environmental crisis with reference to the present time. At the same time, there is a circular temporality, the biological story of continuity and rebirth, staged through a picture of the circle of life, with the frog's different phases: from egg to tadpole, to a frog with tail, to a frog without tail. Added to this is the evolutionary deep time, represented by a frog fossil. The purpose here is not to explain evolution, which would have been a possibility. Instead, the fossil contributes to the understanding of how dramatic the present environmental changes are, signalling that this is not just a change among others but a change of something that has been around for a very long time, and that is happening at an alarming speed. The future is not articulated; instead, the focus is on the ambition to restore the threatened balance (which could be said to indicate a future in balance). In a way, the amphibian exhibit's displays combine orientations and complications. It also hints at evaluations and resolutions in its summary of the general environmental story. Together, these features pave the way for the evaluations and the resolutions that arrive at the end of the path through the aquarium.

The evaluation makes the story

The evaluation states the reason for telling a story (Labov and Waletzky 1967). In the case of the aquarium, the overall evaluation of the story told is connected to its stated mission, which could be simply rendered as to 'change behaviour!' The habitats stage a nature in balance, while the environmental exhibits emphasise change, and together this stages a marine world in danger. I locate the purpose of the story in the relation between these two chronotopes: namely, to keep the balance of the first, we need to alter the second, demanding a change in behaviour.

So, how is this presented at the aquarium? According to Barbara Johnstone, the evaluation spreads out along a storyline until it reaches its most direct statement just before the final result or resolution (2016, 546). The second half of the aquarium's lower level holds several stations corresponding to this description. Placed after the Antarctic and before the North Pacific is an auditorium showing a film about the Oceanário de Lisboa. The film tells a story of its own, but it helps to evaluate the aquarium's overall story as well. The film starts with the earth shown from a distance with dark space as a backdrop, then continues with a sunrise and a voice speaking in Portuguese, subtitled in English. The continued story presents the large aquarium tank and explains how it forms the core of the exhibition, connecting the separated habitats to give the illusion of a single, global ocean and thus stating that there are no frontiers in the ocean, only 'currents, cycles, and ecosystems, fundamental for the balance of the whole planet'. The film ends with a message evaluating the main story told at the aquarium:

> The ocean supports life. From the fish we eat, to the air we breathe, the survival of the Planet Ocean depends on our capacity to protect marine ecosystems and to guarantee their sustainability. Oceanário de Lisboa is this window open to the magic of diversity that aims to excite, educate and inspire visitors, you sitting here, to put into practice a new citizenship of ocean conservation.[8]

The evaluation comes in many forms, but its basic message is simple: change behaviour. So far, the environmental story could be summarised as a staged marine world in peril, in need of balance, and demanding changed behaviour. The next section will develop how this could be done.

The resolution at home

In Labov and Waletzky's theory, the resolution ends the story, telling how things turn out – or in this case, opening for possible ways to slow down environmental changes. This is the situation at the aquarium, where the environmental story implies how the problem of threats to nature can be solved. Although already addressed in the amphibian exhibit, the possible resolution comes at the end, in an exhibit called the House of Vasco that is designed for children and designed around a Fox International Channel television character named Vasco. The Vasco character is a young boy, dressed like a diver, in a blue swimsuit with the Oceanário's logo printed on the chest. Just like the other environmental exhibits, The House of Vasco is not found on the map of the aquarium.

The starting point of the exhibit is framed with illustrations of bookshelves with ocean-themed literature, and I recognise *The Odyssey*, Jules Verne's *20,000 Leagues under the Sea*, Herman Melville's *Moby Dick*, and Ernest Hemingway's *The Old Man and the Sea*. Next to this is a screen showing clips from the television series presenting Vasco. The opening sequence alludes to Superman, another animated character, as a voice asks, 'Is it a dolphin? Is it a shark? A propulsion submarine? No. What is it …'. Vasco is swimming swiftly through the water before introducing himself: 'Hello, I'm Vasco. I was born on the 8th of June, World Ocean Day'. He continues to tell how he likes to dive and to explore the ocean, saying that when he grows up, he wants to be a scientist, a diver, a naval engineer, or maybe an astronaut, a farmer, or an explorer. He adds that he wants to tell us a secret; he is a superhero. 'Yes, I know, I do not have a cape, but what kind of superhero needs a cape under the surface?' A set of clips shows how he saves the ocean by removing plastic debris, ending with a plea: 'Let's help and preserve the oceans'.

There are several things worth noting in Vasco's self-presentation. To start, the introduction to Vasco presents a global chronotope. The name Vasco hints at the Portuguese ocean explorer Vasco da Gama, known for rounding the Cape of Good Hope and finding a sailing route to India. Likewise, Vasco is born on World Ocean Day, a global day dedicated to events highlighting the ocean condition, often aiming at pollution. The home is decorated with well-known books such as Jules Verne's and Hemingway's stories of humans in strained relationships with nature. At the same time, by describing himself as a superhero, Vasco connects himself to a cadre of world saviours. Though starting with worldwide ambitions, Vasco primarily addresses the local and the everyday. The character works as a trickster, a flexible character connecting different worlds and instigating change (Hyde 1998). Traditionally, tricksters were found in religious myths; today they are found in popular-culture narratives, often in films and on television. The trickster of the aquarium's environmental story takes the abstract issue of climate change and locates it in the mundane.

The resolution comes in the form of examples of changes possible to adopt in everyday life, what can be defined as *banal sustainability*, emphasising the capacity for environmental considerations in the everyday and the problems of everyday domestic consumption in order to emphasise a general responsibility for the environment (Kaijser 2019; cf. Billig 1995). Essentially, this resolution is an act of empowerment, allowing everyone to participate in slowing down detrimental environmental change. This means implementing a certain worldview, a way of fostering environmental awareness in everyday life, linking the resolution to the mission of the Lisbon aquarium and their ambition to 'change behaviour'.

Close to the screen showing the Vasco film is a sign inviting visitors to be a special guest in Vasco's house and to see how a superhero lives: 'Follow me and you will be an ally in the mission to save the oceans'. One feature of the trickster is the possibility to cross boundaries. Information and stories presented at the aquarium tend to be performed in a general way, as an anonymous voice representing the aquarium speaking to an anonymous audience. The voice of Vasco was more direct, talking to the visitors, relating scientific insights to a popular understanding and, in doing so, avoiding a scientific idiom. The exhibit displays how a contribution to conservation could be made. The exhibit stages two rooms: a kitchen and a bathroom built at a scale suitable for children. The kitchen features one area with a refrigerator, stuffed with vegetables, fruits and yogurts, while the second has a stove with a soup boiling on the stovetop, with pumpkin, carrots, garlic, spinach, onions, turnips, and potatoes. On the wall, signs explain different ways to reduce the use of water, how to avoid food waste, and recommendations such as 'Most of the energy we use at home comes from sources that, besides polluting the environment, will run out one day. If you use energy from a renewable source, you will help reduce global warming'.

The notion of the trickster also relates to the amphibian exhibit, another place at the aquarium where the connection is made between global threats and a local situation, and how the environment could be saved. Frogs and toads have a certain place in culture. Their capacity to change from one shape to another, from water to land, makes them liminal. One of the signs at the aquarium states 'Amphibians have a special place in various cultures, cherished as a symbol of life and good luck. However, others see them as cursed animals'; another reads 'Seen as princes waiting to be kissed or witchcraft ingredients, toads and frogs are fascinating animals, fragile and in danger'. The frog's role in fairy tales and other stories is sometimes that of a trickster and they are a resource in environment-themed storytelling, and this applies to the frogs at the amphibian exhibit, as well as for the character of Vasco. Compared to the House of Vasco, the signs in the amphibian exhibit differ in scope, addressing both decision-makers, everyday actions, and household consumption when advising decision-makers to increase the supervision and monitoring of watercourses and to the public to watch out when driving for amphibians crossing the road.

The resolution is a way of balancing threats to marine environments. The aquariums tell a story of a marine world in danger, but the resolution shows that there is still hope to make things better. Temporality is also important here, but it has a very special character. The Vasco exhibit promotes recycling and reducing, and in this unites temporalities associated with cyclic temporality and change. In this way, the trickster also connects the two chronotopes discussed above, showing how it is possible to turn global change into local circulation.

Consuming the coda

The coda marks the end of the story, and it creates an opening for others to talk or act. The route along the path tells a story of changing marine environments that is oriented within issues of conservation and biodiversity and that suggests how to contribute to rescuing this world by changing habits. But the story does not end here: it has a coda, which appears as visitors leave the aquarium building with the general environmental story and enter the shop located on land. In Labov and Waletzky's theory, the coda is separated from the rest of the story; at the Oceanário de Lisboa, this is highlighted by the coda being in another building. At the entrance of the shop is a sign:

> #SeaTheFuture.
> We see a world with a future.
> A world filled with responsible choices. With local products and sustainable materials.
> Each choice is an opportunity to make a difference and protect the ocean.
> Join us.
> The future starts here.
> A movement for the future.

Additional signs in the shop inform visitors of environmentally friendly choices: 'We see a future with local products, where each choice is an opportunity to reduce environmental impact. Choose to buy local. Products that travel long distances waste more energy and release more carbon dioxide for the planet'. The coda emphasises the message, but more importantly, it connects the experience of visiting the aquarium, and the general information on how to act in an environmentally friendly way, to the present situation, and the possibility of immediately adhering to the stated resolution and making eco-friendly choices. This is the opportunity to be like Vasco, a superhero, and to reach for sustainable solutions. The aquarium's environmental story is tied to the possibility of a happy ending. But the responsibility of making this happen is left to us – the visitors or basically anyone – who need to change our behaviour to continue a life in balance, or to continue our lives as before, and to meet a changing and unbalanced world.

Conclusion

The importance of keeping the marine environments in balance is the key message at the aquarium. This is staged as an environmental story, performed along the path through the aquarium. The educational ambition relies on a repetition of astonishment at the wonders of the world combined with complicating actions that threaten the natural balance. The habitats recreate generic spaces, not places, with certain characteristics of geology, flora, and fauna attached, dominated by a cyclical temporality, while the exhibits on environmental changes highlight a temporality based on change and human impact. The habitat is a chronotope relying on scientific notions, as is climate change. Together they form the aquaria chronotope, a popular scientific chronotope characterised by synchrony. There is no denying climate change, but neither are there any recommendations for how to adapt to a changing world; the aim is to regain the balance. The organised route stresses the importance of changing everyday routines, a banal sustainability. The story's main ingredient is neither the animals nor the habitats, but a message focussed on conservation.

Having said this, it is optional to engage in the environmental storyline, and this is related to the spatial organisation of the aquarium and the variety of ways to walk the story or, to be precise, the possibility to walk outside of the environmental story. The way that the path is organised makes it impossible to move through the aquarium without experiencing the habitats and its animals. This is not the case with the exhibits dedicated to environmental issues, where it is optional to enter. The One Ocean, One Planet, the amphibian exhibit, and The House of Vasco are placed in the space between the habitats (this goes for the exhibit on ocean plastic as well). If the big tank is placed on the left-hand side, the four environmental exhibits are on the right, and it is possible to follow the path while looking at animals and avoiding the signs and the exhibits between the habitats, thus taking no notice of the environmental message. The facts enumerated in the exhibits that disseminate knowledge of the oceans respond to the first part of the mission of the aquarium. The second part of the mission, to raise awareness and to change behaviour, calls for the route incorporating the exhibits on the right.

Though this chapter concentrates on the environmental story staged at the aquarium, it is worth reflecting on what I could detect from also noticing other visitors at the aquarium. Based on my observations, and this could be coincidental, most visitors missed out on the environmental story, engaged by the large tank while ignoring the environmental exhibits. Even though aquariums generally highlight their environmental mission, it seems to me that most visitors prefer to be mesmerised by the animals' continuous movement in the tanks – as shown, an experience with a chronotope of its own. Considered on its own, the habitat chronotope, with its cyclical time, leans towards a more idealised notion of nature – one that is not in line with the general environmental story told at the aquarium.

The aquarium works as a nexus connecting several areas of interest. As mentioned, it disperses a scientific viewpoint of acidification, rising sea-levels, and global warming into a popularised story of consumer patterns, shifting focus from bars and charts to lifestyle choices. The issue of banal sustainability not only emphasises personal responsibility but also translates environmental concerns into identity projects, with considerations of how to best live an environmentally friendly life. Also important is the way in which the environmental story connects the global to the local, and vice versa. While the notion of climate change entails a global viewpoint, at the aquarium this scope is narrowed and localised to different habitats or reframed as a question of the survival of individual species, making climate change into something geographically defined and detectible. At the same time, the spatial organisation of the aquarium, with its continuous contact with the large tank, underlines the 'one world, one ocean' concept. As all the staged habitats also contain an environmental message, the experience of moving through the aquarium makes it obvious that this one world is made up of a string of habitats, constituted through widely different living conditions, with one thing in common: they are all in peril.

Notes

1 www.oceanario.pt/en/about-us/history/.
2 www.oceanario.pt/en/news/oceanario-de-lisboa-is-the-worlds-best-aquarium.
3 The article is a report from the research project *Staged Nature, Public Aquariums as Institutions of Knowledge*, financed by the Swedish Riksbankens Jubileumsfond. During this project, over 50 aquariums were visited, with documentations of exhibits, habitats, and animals. At fifteen of these aquariums, I have also conducted in-depth interviews and walked the premises with the staff at the aquariums. This is an important foundation for the discussions in this chapter. The visit to the Oceanário de Lisboa took place on 13 April and 14 April, 2019. I am thankful to my colleague Hanna Jansson who gave me important comments on an early draft of the article, and to Ella Kaijser who also helped me with proofreading and encouraging comments.

4 For an extended discussion on the different meaning of 'example', see Bjørnstad (2019); Eriksen (2018).
5 For example, this has been the case at interviews made at Universeum in Gothenburg, the Aquarium of the Pacific in Long Beach, CA, and at the Monterey Bay Aquarium, CA.
6 www.oceanario.pt.
7 There is an exhibit on plastics in the ocean at the Oceanário de Lisboa as well, but I leave that out of the discussion as it falls outside the aim of this chapter.
8 This is a quotation from an untitled film shown at the auditorium. The film was screened in Portuguese with English subtitles; the quotation refers to the subtitles.

References

Bakhtin, Michael. M. 1991. *Det dialogiska ordet.* Gråbo: Anthropos.

Billig, Michael. 1995. *Banal Nationalism.* London: Sage.

Bjørnstad, Hall. 2019. 'Response: Let This Be an Example: Three Remarks on a Thematic Cluster about Climate Change Exemplarity'. *Culture Unbound* 11 (3–4): 415–420.

Brunner, Bernd. (2005) 2011. *The Ocean at Home. An Illustrated History of the Aquarium.* London: Reaktion Books Ltd.

Doordan, Dennis 1992. Nature on Display. In: *Design Quarterly*, vol 155:2 (Spring 1992), pp. 34–36.

Doordan, Dennis. 1995. 'Simulated Seas. Exhibition Design in Contemporary Aquariums'. *Design Issues* 11 (2): 3–10.

Eriksen, Anne. 2018. 'Eksemplarisk usedelighet. Eilert Sundts bruk av eksempler i Om sædeligheds-tilstanden (1857)'. *Tidsskrift for kulturforskning* 17 (1): 25–38.

Gröppel-Wegener, Alke and Jenny Kidd. 2019. *Critical Encounters with Immersive Storytelling.* New York and London: Routledge.

Hyde, Lewis. 1998. *Trickster Makes this World.* New York: North Point Press.

Ingemark, Camilla Asplund. 2016. 'The Chronotope of the Legend in Astrid Lindgren's *Sunnanäng*: Toward an Intergeneric Level of Bakhtinian Chronotopes'. In *Genre-Text-Interpretation: Multidisciplinary Perspectives on Folklore and Beyond*, edited by Koski, Kaarina, Frog, and Ulla Savolainen, 232–250. Helsinki: Suomalaisen Kirjallisuuden Seura.

Johnstone, Barbara. 2016. '"Oral Versions of Personal Experience": Labovian Narrative Analysis and Its Uptake'. *Journal of Sociolinguistics* 20 (4): 542–560.

Kaijser, Lars. 2019: 'Promoting Environmental Awareness: On Emotions, Storytelling, and Banal Sustainability in a Staged Rainforest'. *Ethnologia Europaea* 49 (1): 74–90.

Karydis, Michael. 2011. 'Organizing a Public Aquarium: Objectives, Design, Operation, and Missions. A Review'. *Global NEST Journal* 13 (4): 369–384.

Labov, William and Joshua Waletzky. 1967. 'Oral Versions of Personal Experience'. In *Essays on the Verbal Arts. Proceedings of the 1966 Annual Spring Meeting. American Ethnological Society*, edited by June Helm, 12–44. Seattle: University of Washington Press.

Oceanário de Lisboa. 2020. www.oceanario.pt; https://www.oceanario.pt/en/about-us/history/; https://www.oceanario.pt/en/news/oceanario-de-lisboa-is-the-worlds-best-aquarium; https://www.oceanario.pt/en/vasco_en.

Palmenfelt, Ulf. 2017. *Berättade gemenskaper. Individuella livshistorier och kollektiva tankefigurer.* Stockholm: Carlssons bokförlag.

Sagan, Carl. 1994. *Pale Blue Dot: A Vision of the Human Future in Space.* New York: Ballantine Books.

White, Hayden 1980. 'The Value of Narrativity in the Representation of Reality'. *Critical Inquiry* 7 (1): 5–27.

Part 3

Cultural histories of climate change temporality

8 The sixth extinction

Naming time in a new way

Marit Ruge Bjærke

Introduction

In 2014, science journalist Elisabeth Kolbert published *The Sixth Extinction: An Unnatural History*. The book appeared on the *New York Times* list of non-fiction bestsellers and won the 2015 Pulitzer Prize in the General Nonfiction category (*The New York Times* 2014; The Pulitzer Prizes 2015). The title refers to the claim that only five mass extinction events have been identified in the timespan of geological history and that we might now be in the midst of a sixth one. The concept 'the sixth extinction' was not something Kolbert had come up with herself, however. She writes that she was introduced to the concept through an article on threatened amphibia in the early 2000s (Kolbert 2014, 6; but see also Kolbert 2009).

English scholar Jeremy Davies (2016) has noted that the presentation of environmental problems in mass media is often based on comparisons between long, geological timescales and the present state of affairs. In the climate discourse, the connection between the present and geological history has been most forcefully established through the use of the concept 'Anthropocene'. The Anthropocene was first suggested by Paul Crutzen and Eugene Stoermer as a name of a new geological epoch, which was to succeed the ongoing epoch, Holocene (Crutzen and Stoermer 2000). Their contention was that the impacts from human activities on Earth and its atmosphere, and especially the effects of greenhouse gases, had now reached such a level as to indicate that the Holocene epoch should be considered at an end. In the discourse on species extinctions, 'the sixth extinction' has a similar function as the Anthropocene. It draws on a geological timescale and invokes the geological concept of mass extinctions to describe an environmental issue. However, while the Anthropocene concept has been closely studied and theorised in terms of its temporalities (Bonneuil and Fressoz 2016; Fagan 2019; Nordblad 2021), there have been few discussions of the concept of 'the sixth extinction' (however, see Rose, van Dooren, and Chrulew 2017, 6).

At the moment, the sixth extinction is everywhere – in scientific publications, in scientists' warnings, in activists' statements, and in the key messages of UN reports (Barnosky et al. 2011; Ripple et al. 2017; Extinction Rebellion

2019; UN Environment 2019). In these different discourses, 'the sixth extinction' works as a scientific concept, as a descriptive term, as a rhetorical device, and as a concept naming a historical time. In this chapter, however, I explore 'the sixth extinction' as a temporal marker. In the geological community, questions such as what a mass extinction is, how rare it is, and how long it may last, have more or less specific, consensus-based answers. As the concept moves between the natural sciences and other areas of discourse such as popular science and politics, however, the understanding of its temporal aspects do not stay unchanged. Rather, the temporalities of the concept take on new meanings and have different implications depending on the genres it enters. My main objective is to investigate what happens when geological timescales meet with other timescales, such as the political. To approach this question, I investigate the temporalities at play when 'the sixth extinction' or 'the sixth mass extinction' is used in different kinds of popularised science. As intermediaries between the natural sciences, the mass media, and the political sphere, popular science texts are places where concepts and temporal understandings from both the natural sciences and the public meet, intertwine, and circulate (Secord 2004).

The chapter starts with an account of what the sixth extinction typically refers to, and how the concept came into use during the 1980s. I then turn to the understandings of temporality and temporal scales implied in the sixth extinction in popular science. The focal point of my analysis is Kolbert's popular science book *The Sixth Extinction: An Unnatural History* from 2014. I expand the analysis of Kolbert's text by comparing it with two other texts on the same subject, published in 1983 and 1995. The text from 1995 is a popular science book with the same main title as Kolbert's: *The Sixth Extinction: Patterns of Life and the Future of Humankind,* written by paleoanthropologist and conservationist Richard Leakey and science journalist Roger Lewin. The other text, from 1983, is a report from a scientific meeting entitled 'No dinosaurs this time', written by the same Roger Lewin for the scientific journal *Science.* The three texts can all be understood as natural science popularisations, although they represent different levels of popularisation: the text from 1983 is a popularisation within the scientific community, while the two others address the general public. I end the chapter with a comparison between the sixth extinction and the Anthropocene, and a discussion of the political implications of 'the sixth extinction' as a concept.

The size of an environmental problem

Elizabeth Kolbert is not a scientist herself. Her take on the sixth extinction in *The Sixth Extinction: An Unnatural History* is a journalistic one. In each chapter, she presents a different case about a species or a group of species that is either already extinct or threatened with extinction. Kolbert conducts field trips, visits research stations, and goes to zoos to do interviews with scientists working with these species. The starting point of her book, however, is the idea of a sixth extinction, which she describes in the prologue of the book:

> Very, very occasionally in the distant past, the planet has undergone change so wrenching that the diversity of life has plummeted. Five of these ancient events were catastrophic enough that they're put in their own category: the so-called Big Five. In what seems like a fantastic coincidence, but is probably no coincidence at all, the history of these events is recovered just as people come to realize that they are causing another one. When it is still too early to say whether it will reach the proportions of the Big Five, it becomes known as the Sixth Extinction (Kolbert 2014, 2).

The five catastrophic events that Kolbert refers to are so-called mass extinctions. Mass extinctions are defined by geologists as periods in the geological history when extinction rates are so much higher than the speciation rates that a large percentage of the species existing on Earth disappears at approximately the same time.[1] According to Barnosky et al. (2011), mass extinctions are defined by a relative acceleration of extinction rates relative to speciation rates, so that 'over 75% of species disappear within a geologically short interval – typically less than 2 million years, in some cases much less'.

The five mass extinction events which are termed 'the Big Five', and which serve as the background for the idea of an ongoing sixth extinction event, were identified by Raup and Sepkoski (Raup and Sepkoski 1982; Hallam and Wignall 1997, 3). These are the End-Ordovician extinction (about 450 million years ago), the late Devonian extinction (about 375 million years ago), the End-Permian extinction (about 250 million years ago), the End-Triassic extinction (about 200 million years ago), and the End-Cretaceous extinction (about 66 million years ago). Except for the End-Cretaceous extinction, in which, most scientists seem to agree, an asteroid impact played an important part, there is still scientific debate about the cause or causes of each of the Big Five mass extinctions. Some of the explanations are massive volcanism, drops in sea level, global warming, or global cooling.

When the ongoing species extinctions are merged into one event, termed 'the sixth extinction', several things happen. First, species extinctions caused by humans are seen as *one* geological event rather than as a number of separate instances or a historical event. Second, this event is linked to an existing series of five geological mass extinctions and is gradually understood as part of this series of events. The link between the event and the series of mass extinctions indicates that the ongoing extinction will be of about the same size as the previous five mass extinctions. Third, the ongoing species extinction event is placed on a geological timescale. The fact that there have been only five similar events in a very long period of time shows the reader how rare the event is. In this way, the concept of the sixth extinction as a temporal marker couples ongoing species extinctions to ideas of magnitude and rarity, and thus highlights the severity of the matter at hand. It is what Kolbert herself terms 'mind-boggling' (Kolbert 2014, 7).

As stated above, the five extinction events that have later been termed the 'Big Five' were identified in 1982 (Raup and Sepkoski 1982). During the early 1980s, mass extinctions were a hot topic in the geological community. There were debates concerned with, for instance, whether mass extinctions were cyclic or stochastic, whether they were mostly catastrophic events or drawn out in time, and whether the extinction of the dinosaurs had been caused by a massive asteroid impact or not (see, for instance, Benton 1985, 496–497; Lewin 1985, 640).

Several researchers in the late 1970s and early 1980s had argued that the ongoing species extinctions in the late twentieth century were of a massive size, even on a geological timescale (see, for instance, Myers 1980, 101; Eisner et al. 1981). Still, an article in *Science* from 1983 entitled 'No dinosaurs this time' seems to have been the first article within the scientific discourse in which the environmental problem of species extinctions was presented as a new mass extinction. In this article, science journalist Roger Lewin reported from a meeting on the dynamics of extinction in Flagstaff, Arizona. Both geologists and biologists were present when ecologist Paul Ehrlich of Stanford University, according to Lewin, stated that 'Earth's biota now appears to be entering an era of extinctions that may rival or surpass in scale that which occurred at the end of the Cretaceous, some 65 million years ago'.

Neither Ehrlich, nor Daniel Simberloff, the other ecologist that is cited in this article, use the concept 'the sixth extinction'. However, they both compare ongoing species extinctions to earlier mass extinction events. Simberloff even contends that the new mass extinction is bigger than the one in the Late Cretaceous and, according to Lewin, states that '[a]ll told, it is clear that the catastrophe we are facing is *not* the worst biological debacle since life began – the Late Permian extinction must be that – but it certainly vies for second place' (Lewin 1983, 1169). Thus, the ongoing extinction of species is placed in a hierarchy among the mass extinctions that have happened through geological time – not the largest, but larger than the one that exterminated the dinosaurs.

During the next few years following this publication, the connection between the ongoing species extinctions and geological mass extinctions was quickly taken up by conservation biologists. A report from a symposium held in 1984 by the Species Survival Commission of the International Union for the Conservation of Nature and Natural Resources (IUCN), entitled *The Road to Extinction,* shows that the members of IUCN working with species extinctions paid close attention to the mass extinction debates in the geological community. While one participant refers to geologist David Raup's work on cyclic mass extinctions, another participant draws a parallel between mass extinctions today and the mass extinction during the Cretaceous-Tertiary boundary: 'The parallel with the collapse of complex ecosystems suggests that if we do nothing about these collapses, the already predicted mass extinctions may occur' (Fitter and Fitter 1987, 58). In a guest essay in the 1988 *IUCN Red List of Threatened Animals*, Bruce A. Wilcox, Director for the Centre for Conservation Biology at Stanford University, wrote:

> We are entering a period of Earth's history when the global environment is changing on a scale and with an intensity that has occurred only a few times during the nearly four billion-year span of the existence of life on Earth. In the past, such changes were brought about by, for example, gradual tectonic movements in which entire continents repeatedly divided and coalesced, effecting global geographical and climatic conditions. [...] The present period of change, however, which will undoubtedly warrant its own geological epithet by future geologists, is being brought about neither by uncontrollable nor distant terrestrial or extraterrestrial forces, but by us.
>
> (Wilcox 1988, v)

Thus, during the 1980s, mass extinctions became part of the language of conservation and a dramatic new way of presenting the environmental problem of species extinctions in policy-oriented texts, while the temporalities and causes of mass extinctions were still actively discussed in the geological community. The environmental problem of species extinctions became part of a series of events, which had already shown themselves to result in dramatic outcomes, but of which there was still a lot one did not know.

Ehrlich and Simberloff did not construct the new mass extinction as 'the sixth'. For them it could rather be said to be the new number two. They focussed on a hierarchy based on magnitude – the percentage of species being wiped out in each mass extinction event – where the present mass extinction would rank higher than the one in the late Cretaceous, but below the Late Permian one. Neither did Les Kaufman and Kenneth Mallory, who edited the book *The Last Extinction* from 1986 (Kaufman and Mallory 1986). The title carries some connotations of the chronological numbering of mass extinctions but refers more directly to eschatological ideas of it being the last one for humans to experience. However, in the early 1990s, the new mass extinction was firmly established as 'the sixth'. First, ecologist E. O. Wilson used the concept in his book *The Diversity of Life* published in 1992 (Wilson 2001, 30), and then Richard Leakey and the aforementioned Roger Lewin used it as the title of their book *The Sixth Extinction: Patterns of Life and the Future of Humankind* from 1995. Thus, from the middle of the 1990s, the idea of the ongoing species extinctions as the sixth in a chronological series of mass extinctions, separated by large intervals of time within the planet's geological history, was established both in the scientific community and among conservationists, and was in use in popular science books.

In the 1960s, species threatened with extinction had been turned into a global phenomenon encompassing all species groups and all parts of the world. Important in this process was the idea of a global environment emerging in the first half of the twentieth century, and technologies such as the construction of Red Lists and Red Data Books of threatened species by the IUCN (Warde, Robin, and Sörlin 2018, 35–41; Bjærke in press). In the 1980s, the comparison between mass extinctions from geological history and the ongoing species extinctions highlighted that the phenomenon was not only large on a spatial scale. It was standing out on a similarly large temporal scale, encompassing the complete history of life on Earth. Thus, the new idea of an ongoing mass extinction showed the problem of species extinctions to be even more dramatic than before. The sixth extinction was of a global nature, both when considered in space and in time.

The history of a species

Already in 1983, however, the new mass extinction was not defined only by its similarity to earlier ones. In the article from the Flagstaff meeting, Ehrlich is reported to have said, 'The current extinction is, however, different [...]: For the first time in geologic history, a major extinction episode will be entrained by a global overshoot of carrying capacity by a single species – *Homo sapiens*'. Thus, while the ongoing mass extinction is considered similar enough to the mass extinctions of the geologic history to be included in a series, it is also different and stands out, since it is caused by the human species. This view of the new mass extinction as both similar and completely different from the past is typical for all three texts that I close-read in this chapter. Leakey and Lewin (1995, 241) state that '*Homo sapiens* is poised to become the greatest catastrophic agent since a giant asteroid collided with the Earth sixty-five million years ago', while Elizabeth Kolbert starts her story of the sixth extinction with the following:

> Beginnings, it's said, are apt to be shadowy. So it is with this story, which starts with the emergence of a new species maybe two hundred thousand years ago. The species does not yet have a name – nothing does – but it has the capacity to name things.
>
> (Kolbert 2014, 1)

Kolbert continues her story in the same manner. She states that the individuals of the new species are resourceful, and describes how they travel to new regions, hunt, interbreed with similar species, kill off these similar species, cross the seas, increase in numbers, raze forests, and how, in their wake, other species are 'wiped out'. Although she includes some historical events, such as the finding of oil and gas, there are no individual humans in Kolbert's story. It starts as far back as 200,000 years ago with the emergence of a new species – humans – and ends at some undefined point of time in the future.

Kolbert makes naming things the main characteristic of the human species. When it first emerges it has 'the capacity to name things'. Then, when the species starts to reproduce at an unprecedented rate, she adopts the term that it has created: '*Homo sapiens,* as it has come to call itself'. Towards the end of the book, Kolbert couples the ability to name things directly with the extinction of other species: 'With the capacity to represent the world in signs and symbols comes the capacity to change it, which, as it happens, is also the capacity to destroy it' (Kolbert 2014, 258).

Kolbert's story of the sixth extinction not only brings in humans as the cause of the sixth extinction, it also focusses on the human as a species, with certain traits that do not change over time. The way Kolbert uses the species concept and also includes the Latin epithet for the human species, *Homo sapiens*, in her text points clearly to the natural science discourse on which her presentation of the sixth extinction is based. The sixth extinction is not about individuals, individual choices, specific nations or cultures, but about a species. The human in Kolbert's story is static and unchanging. It acts the same way towards other species from the beginning of her book to the end. 'Though it might be nice to imagine there once was a time when man lived in harmony with nature, it's not clear that he ever really did', Kolbert concludes (2014, 235). Thus, she conveys the understanding that the human species, because of its inborn traits, seems to be bound from the beginning to cause a sixth extinction.

The presentation of the human species with inborn traits that lead to mass extinction shows that, in addition to the geological timescale of life on Earth, an evolutionary timescale plays an important part in Kolbert's presentation of the sixth extinction, while the historical timescale of periods and cultures is almost completely absent. A similar temporal structure is present in Leakey and Lewin's book from 1995. They also argue that the human species have a basic capacity to cause a sixth mass extinction. '[W]e arrived equipped with the capacity to devastate that diversity wherever human populations traveled', they state (Leakey and Lewin 1995, 233). In the report from the Flagstaff meeting in 1983, however, the capacity to cause a mass extinction is not presented as an innate trait in the human species. Here, Ehrlich instead links the idea that the human species are causing a mass extinction to recent developments, such as urban growth, chemical pollution, felling of tropical forests, and the present size of the human population. Ehrlich was himself the author of the book *The Population Bomb* from 1968, in which he stated that '[i]t is fair to say that the environment of every organism, human and nonhuman, on the face of the Earth has been influenced by the population explosion of *Homo sapiens*' (Ehrlich 1968, 46). Thus, to Ehrlich it is not the human species itself that causes the mass extinction but the present population size of the species.

When contrasting these statements, it is clear that in Leakey and Lewin and Kolbert the connection between the mass extinction and the species *Homo sapiens* has turned into something different and more fundamental than in Lewin's text from 1983. An evolutionary timescale of innate species traits has replaced the ecological timescale of population dynamics, and the extinction of other species has become intrinsic to the human species. Therefore, it is not the use of a geological timescale in itself that leads to such a determinism, but the two-step process through which the sixth extinction is based on a combination of the geological and evolutionary timescales. Together, the two timescales act as the basis for presenting human history as species history, removing the sixth mass extinction from the importance of specific human actions, historic periods, or civilisations, such as 'industrialisation' or 'the Western world' through an evolutionary temporality that we might call *Homo sapiens* time.

From habitat destruction to climate change

Today, climate change seems to have become *the* environmental problem that includes other environmental problems such as biodiversity loss (Hulme 2011; Bjærke 2019). The idea that present species extinctions could be seen as a new mass extinction of species caused by humans did, however, emerge and evolve independently of the emergence of climate change as an environmental problem in the 1980s and 1990s. As cited earlier, the article from the meeting where Ehrlich and Simberloff first presented the human species as the cause of a mass extinction attributed the mass extinction to several concrete factors, such as urban growth, chemical pollution, and agricultural development. However, it cites Ehrlich to have claimed that the felling of tropical moist forests is 'the single most extensive potential impact on future species diversity' (Lewin 1983, 1168). Leakey and Lewin, in their book from 1995, also connect the sixth extinction with tropical forests. According to them, the three principal ways in which humans endanger the existence of other species are (over)exploitation, the introduction of alien species, and the destruction and fragmentation of habitat (Leakey and Lewin 1995, 234). Like Ehrlich and Simberloff, Leakey and Lewin consider the clearing of tropical rainforests to be the single most important of these, and devote several pages to a discussion of estimates of the number of species extinctions based on estimated reductions of tropical rainforests (Leakey and Lewin 1995, 236–245).

Leakey and Lewin's book seems to have been influenced rather directly by Ehrlich's description of the problem of species extinctions, as they refer to statements from Ehrlich and E. O. Wilson at a meeting in 1986, about which Lewin also wrote a news article in *Science* (Lewin 1986). In the middle of the 1990s, when Leakey and Lewin's book was published, climate change was, however, already a topic considered at least partly related to species extinctions. In a review of Leakey and Lewin's book, Thomas E. Lovejoy wrote, 'It is a compelling introduction [...] to the reasons why the sixth extinction is generically different from previous mass extinctions – and certainly not in humanity's interest. Regrettably, there is no mention of the implications of human-induced climate change [...]' (Lovejoy 1996, 594).

In Kolbert's book, the description of the cause of the sixth extinction has changed. In the aforementioned prologue, she writes:

> Meanwhile, an even stranger and more radical transformation is under way. Having discovered subterranean reserves of energy, humans begin

> to change the composition of the atmosphere. This, in turn, alters the climate and the chemistry of the oceans. Some plants and animals adjust by moving. They climb mountains and migrate towards the poles. But a great many – at first hundreds, then thousands, and finally perhaps millions – find themselves marooned. Extinction rates soar, and the texture of life changes.
>
> (Kolbert 2014, 2)

According to Kolbert, then, climate change is radically different from hunting, the crossing of the sea, and the razing of forests. Changing the atmosphere is presented as the final and major step in the story – the phenomenon that leads to the fact that 'extinction rates soar, and the texture of life changes'. Thus, in Kolbert's book, although the sixth extinction starts with the emergence of the human species, the major driver of the sixth extinction is climate change. The connections with population growth and with the clearing of tropical rainforests are played down.

The expectation of the duration of the sixth extinction also varies between the three different texts. I have already shown how Leakey and Lewin, as well as Kolbert, expand it to encompass the whole history of the human species, 200,000 years back in time. However, the three texts also have slightly different ways of including the future in their discussions. According to Lewin (1983), Ehrlich addressed the geologists at the conference in Flagstaff with the contention that the new mass extinction has a much shorter timespan than the earlier mass extinctions. Lewin writes, 'Within perhaps 100 years – 200 at most – the total of recently extinct species in temperate and tropical regions will likely equal the casualty toll of the event 65 million years ago' (1983, 1168). The article underlines that this time frame is so short compared to earlier mass extinctions that palaeontologists and geologists are almost unable to grasp it (Lewin 1983, 1168).

In their 1995 book, Leakey and Lewin describe the sixth extinction in the fourth part, entitled 'The Future', in which they discuss the evidence of whether there is an ongoing mass extinction of species or not (Leakey and Lewin 1995, 232–245). Scientific in scope as the book is, Leakey and Lewin present and discuss the rate of felling of tropical forests, Wilson's model of island biogeography, current estimates of the number of existing species, and the consequent estimates of the current extinction rate before they reach their conclusion that early in the next century, extinction rates will be comparable to the five earlier mass extinctions known from the geological records if current trends of habitat destruction continue. Thus, the sixth extinction is a result of current trends in habitat destruction (especially the clearing of tropical rainforests), but is placed in the near future – more specifically in the early twenty-first century (Leakey and Lewin 1995, 242).

Kolbert's book is less scientific in scope than Leakey and Lewin's, and she does not present precise estimates of either present or future extinction rates, or of when exactly the mass extinction will have happened. Kolbert instead moves almost seamlessly from the past to the present and then to the future in her presentation of the history of the human species and its extermination of other species. The part of the prologue in which she presents her story of *Homo sapiens* is all written in present tense, and the active blurring of the relationship between the past, the present, and the future produces the feeling that the future is just as well known as the present. The only indication that some part of Kolbert's sixth extinction has not happened yet, and that she does not know exactly how it is going to be, is the word 'perhaps' in the phrasing 'finally perhaps millions [of species] – find themselves marooned' (Kolbert 2014, 2). Kolbert's text, with its reference to climate change as the ultimate cause of the sixth mass extinction and its blurring of the relationship between present and future, creates a temporal uncertainty that stretches the timescale of the sixth extinction further into the future than the earlier texts, but at the same time also places its cause more firmly in the present.

The main reason for the openness of the sixth extinction as a concept is the many different timescales that are incorporated into it. As a geological event, it is part of the timescale of the history of life on Earth. When including the human *as* species, it draws either on ecological temporalities connected with population dynamics and exponential growth or on the evolutionary timescale of a species, with its specific inherent and unchanging characteristics that separate it from other species. As a popular and political concept coupled with climate change, however, the sixth extinction also draws on the various timescales connected with climate change, such as catastrophe and crisis.

The instability that results from these different timescales is enhanced by the fact that the timescales of mass extinctions as geological events are subject to ongoing discussions within the geological community. The geological fossil record does not usually offer the necessary resolution to determine whether a mass extinction happened in 10, 10,000, or a 100,000 years (Barnosky et al. 2011), and research on catastrophic and rapid events in the geologic past is an active research area. When the ongoing species extinctions are turned into a geological mass extinction event, the undecidability of the geological moment thus enters into the historical and social temporalities of environmental debate in a triple fashion: because deep time in itself is difficult to grasp if one is not a geologist or a natural scientist; because geology itself does not provide clear answers to how quickly a mass extinction can really happen; and because such long timespans are difficult to align with political and historical temporalities.

The sixth extinction and the Anthropocene: systemic concepts and politics

Both the Anthropocene and the sixth extinction are concepts created during the last 50 years to describe certain ideas about how the present connects with the past. They are obtained from the geological sciences, and derive much of their strength from the conception that the human species wields a different kind of force than it did before, which, according to various researchers, calls for a connection between historical and geological time (Robin and Steffen 2007; Chakrabarty 2009; 2018). While the Anthropocene was a chronological concept from the start, proposed as a new epoch on the stratigraphic timescale, the idea of an ongoing mass extinction did not become a chronological concept until it was numbered as the sixth in a chronological series of mass extinctions starting with the End-Ordovician extinction. As chronological concepts, however, they both place present human actions on a geological timescale: the Anthropocene concept by inventing a new name for the present geological epoch, and the sixth extinction by describing the present as the sixth in a series of rare events of large geological significance.

The two concepts are closely intertwined scientifically. As stratigraphic units or time periods are separated by changes in the fossil records in addition to other kinds of evidence, extinction events are typically border events between epochs, periods, or eras. Thus, evidence of a sixth mass extinction would be one of the ways in which a future geologist would demarcate the Anthropocene (Zalasiewicz et al. 2017). From Wilcox's essay from 1988, cited earlier in this chapter, it is also possible to see how the discussions of the new mass extinction brought with them an expectation of a new geological name for the present: 'The present period of change [...], which will undoubtedly warrant its own geological epithet by future geologists'. Thus, the idea of an ongoing mass extinction is both a prerequisite for and incorporated in the Anthropocene as a geological concept.

As temporal markers used in popular discourses, however, the sixth extinction and the Anthropocene function in very similar ways. They are both systemic concepts based on a geological timescale, they both present humans as a geological force, and they both describe present understandings of environmental problems as a recent awakening, fronted by natural scientists (Steffen, Crutzen, and McNeill 2007, 614; Kolbert 2014, 3). Even the tales of how the concepts were first proposed are very similar. The Anthropocene is told to have been proposed by Paul Crutzen at a conference under the International Geosphere-Biosphere Programme in 2000 (Warde, Robin and Sörlin 2018, 165). The first proposition of the ongoing species extinctions as a mass extinction comparable to the large mass extinctions of geological history also seems to have happened at a conference. Both persons proposing the new concepts were natural scientists but not geologists. The scientific lecture for a scientific audience is a very specific timespace, which allows both for connections between the different natural sciences and for propositions and concepts of a slightly more popular character than the scientific article. Thus, the scientific conference speech seems to be an apt genre for introducing concepts that are easily adapted into discourses of politics and environmental activism.

Concepts such as the sixth extinction and the Anthropocene highlight the close connection between natural sciences, science popularisations, and environmental rhetoric and politics. Both the sixth extinction and the

Anthropocene belong to a cluster of concepts propelling the large, systemic narratives from the natural sciences into the domain of human affairs and political discussions. They originated within the natural sciences, but the way they were proposed – at conferences and by non-geologists – show that they were never meant as purely scientific concepts (if there ever was such a thing).

The Anthropocene concept has been criticised for its political and historical implications as a grand narrative. For instance, human ecologists Andreas Malm and Alf Hornborg (2014) have claimed that the concept forces a certain kind of historical understanding, namely species history, and blames climate change on the human species rather than on the relatively few people and societies that have actually benefitted from fossil fuel technologies. It has also been claimed that the Anthropocene, with its focus on Earth systems, implies a very simplistic representation of the history of present environmental problems, presenting it as a gradual awakening rather than as a result of various actors, intentions, or historical discontinuities in the long history of environmental thinking (Bonneuil and Fressoz 2016). Based on such criticisms, the use of the concept in the political discourse has been criticised for appearing deterministic and for producing few perspectives on how environmental problems can be met and discussed politically (Ekström and Svensen 2014; Moore 2016; Nordblad 2021). The idea that we are part of the Anthropocene, a geological epoch that has not yet ended, seems to imply that it is declared from an imagined point in the future, millions of years from now, and thus turns the present into an imagined and already determined past (Nordblad 2021). The present is seen from such a distance and on such a scale that the details disappear, what is at stake becomes hidden, and the possibility of change seems small.

As I have shown in my analysis, the concept of the sixth extinction is an even more obvious example of species history than the Anthropocene concept, as it draws heavily on the species concept and on evolutionary time. As the sixth extinction is also a systemic concept from the natural sciences, arguments very similar to those concerning the Anthropocene can be made regarding it as a political concept. It seems to be designated from a distant future, and its long temporal scale blurs the various political discussions and measures that are necessary to do something about the problem into one large, geological event. This is clearly seen in the way the understanding of the cause of the sixth extinction has gradually changed during the last 30 years, from the felling of tropical forests to climate change, with few other changes in the use of the concept itself.

However, although historian of ideas Julia Nordblad's argument that the Anthropocene appears to be declared from an imagined point in the future also holds for the sixth extinction, I rather think that the most important temporal feature of the sixth extinction is the way it is used to combine this geological timescale with a starting point that coincides with the emergence of the human species. Political science scholar Madeleine Fagan (2019, 62) has discussed how different choices of starting date for the Anthropocene includes or excludes different groups of people as they highlight different parts of and different causes for present environmental problems. She has also shown how discussions of such different starting dates might be one way of destabilising linear and exclusionary accounts of the human as a planetary force (Fagan 2019, 62). The sixth extinction, in the presentations of Kolbert as well as Leakey and Lewin, only offers one starting point: the emergence of the human species. The capacity to cause the sixth extinction is presented as integral to human nature, making it seem inescapable. Thus, although the Anthropocene bears the name of the human – the Anthropos – the sixth extinction as a concept relies even more heavily on the species view of humanity than the Anthropocene. The human species is the culprit in the story, and in such an account there is no room for the smaller things such as nation states, ideologies, or variations in ways of living. Thus, it is not mainly the fact that the sixth extinction is 'declared from the future' that leads to its determinism and its unsuitability as a political concept; rather, it is the fact that it is being 'declared from the past' – as an inherent part of being human.

Concluding remarks

My analysis has concentrated on three aspects of the sixth extinction as a temporal marker. First, that the texts using the concept depict the ongoing species extinctions as a global environmental problem and place them on a geological timescale, and that as global and geological phenomena mass extinctions are extremely rare. Second, that the texts that use the concept 'the sixth extinction' depict human history as a species history, not as a history of individuals, groups, or nation states, through the combination of a geological and an evolutionary timescale. The sixth extinction is depicted as innate to the human species. Third, and partly because of the way the human species is perceived as predetermined to cause a sixth extinction, that the concept is open to the introduction of new causes of extinction and changing expectations of its time frame. During the three to four decades that the idea of a sixth mass extinction has existed, both the time frame of the extinction event and the perceived main cause of the extinction have altered significantly – from being caused by the ongoing clearing of tropical rainforests, to being the result of an expected climate change catastrophe.

Just as with the Anthropocene concept, the temporalities of 'the sixth extinction' makes it a problematic concept when used in political discourse. As used by Leakey and Lewin, and by Kolbert, the concept introduces a species view of humanity which is difficult to reconcile with the possibility of fixing the problem. The evolutionary temporality of 'the sixth extinction' – *Homo sapiens* time – leaves the concept in danger of obscuring the many political discussions and choices that are needed to actually handle this environmental problem, turning them into a question of the inherent nature of the human species, and a geological event that is already out there.

With the dramatic connotations of its focus on the large size and extreme rarity of the ongoing species extinctions, 'the sixth extinction' obviously has an important rhetorical function. By comparing the present species extinctions with something that has only happened a few times in the history of the Earth, it highlights the size of the problem and can thus work as a moral and political incitement. Still, to be able do something about species extinctions, it is necessary to confront a range of different causes and pressures, solutions, and measures, which have different consequences for different people and which are varying and changing on temporal and spatial scales different from those on which the sixth extinction is based. Politicians, scientists, and environmental organisations using the concept to highlight ongoing species extinctions therefore need to be aware that the consequences of using the concept might be that the time for political action is replaced with a *Homo sapiens* time, characterised by words such as 'inescapable' and 'too late'. As a temporal marker, 'the sixth extinction' draws on completely different timescales than those on which possible solutions to the problem are to be found.

Note

1 Approximately the same time, being in this case on a geological timescale, restricted by the degree of resolution provided by the geological record (Hallam and Wignall 1997, 1).

References

Barnosky, Anthony D, Nicholas Matzke, Susumu Tomiya, Guinevere O. U. Wogan, Brian Swartz, Tiago B. Quental, Charles Marshall, et al. 2011. 'Has the Earth's Sixth Mass Extinction Already Arrived?' *Nature* 471: 51–57.

Benton, Michael J. 1985. 'Interpretations of Mass Extinction'. *Nature* 314: 496–497.

Bjærke, Marit Ruge. 2019. 'Making Invisible Changes Visible: Animal Examples and the Communication of Biodiversity Loss'. *Culture Unbound: Journal of Current Cultural Research* 11: 394–414.

———. In press. 'Little Red Ring Binders: Early Red List Temporalities'. In *Times of History, Times of Nature: Temporalization and the Limits of Modern Knowledge,* edited by Anders Ekström and Staffan Bergwik. New York: Berghahn Books.

Bonneuil, Christophe, and Jean-Baptiste Fressoz. 2016. *The Shock of the Anthropocene.* London and New York: Verso.

Chakrabarty, Dipesh. 2009. 'The Climate of History: Four Theses'. *Critical Inquiry* 35 (2): 197–222.

———. 2018. 'Anthropocene Time'. *History and Theory* 57 (1): 5–32.

Crutzen, Paul J., and Eugene F. Stoermer. 2000. 'The "Anthropocene"'. *The International Geosphere–Biosphere Programme (IGBP) Global Change Newsletter* 41: 17–18.

Davies, Jeremy. 2016. *The Birth of the Anthropocene.* Oakland: University of California Press.

Ehrlich, Paul R. 1968. *The Population Bomb.* New York: Ballantine Books.

Eisner, Thomas, Hans Eisner, Jerrold Meinwald, Carl Sagan, Charles Walcott, Ernst Mayr, Edward O. Wilson et al. 1981. 'Conservation of Tropical Forests'. *Science* 213: 1314.

Ekström, Anders, and Henrik H. Svensen. 2014. 'Naturkatastrofer i menneskets tidsalder: Mot en tverrfaglig forståelse av antropocen-begrepet'. *Tidsskrift for Kulturforskning* 13 (3): 6–21.

Extinction Rebellion. 2019. 'Extinction Rebellion: The Truth'. https://rebellion.earth/the-truth/, accessed June 20, 2019.

Fagan, Madeleine. 2019. 'On the Dangers of an Anthropocene Epoch: Geological Time, Political Time and Post-human Politics'. *Political Geography* 70: 55–63.

Fitter Richard, and Maisie Fitter, eds. 1987. *The Road to Extinction*. Cambridge: IUCN.

Hallam, Anthony, and Paul B. Wignall. 1997. *Mass Extinctions and their Aftermath*. Oxford: Oxford University Press.

Hulme, Mike. 2011. 'Reducing the Future to Climate. A Story of Climate Determinism and Reductionism'. *Osiris* 26 (1): 245–266.

Kaufman, Les, and Kenneth Mallory, eds. 1986. *The Last Extinction*. Cambridge, MA: The MIT Press.

Kolbert, Elizabeth. 2009. 'The Sixth Extinction?' *The New Yorker*, May 18, 2009. https://www.newyorker.com/magazine/2009/05/25/the-sixth-extinction

———. 2014. *The Sixth Extinction: An Unnatural History*. New York: Picador.

Leakey, Richard, and Roger Lewin. 1995. *The Sixth Extinction: Patterns of Life and the Future of Humankind*. New York: Anchor Books.

Lewin, Roger. 1983. 'No Dinosaurs This Time'. *Science* 221: 1168–1169.

———. 1985. 'Catastrophism Not Yet Dead'. *Science* 229: 640.

———. 1986. 'News and Comment: A Mass Extinction without Asteroids'. *Science* 234: 14–15.

Lovejoy, Thomas E. 1996. 'Book Review: How Much Is an Elephant Worth?' *Nature* 382: 594.

Malm, Andreas, and Alf Hornborg. 2014. 'The Geology of Mankind? A Critique of the Anthropocene Narrative'. *The Anthropocene Review* 1 (1): 62–69.

Moore, Jason W., ed. 2016. *Anthropocene or Capitalocene: Nature, History, and the Crisis of Capitalism*. Oakland, CA: PM Press.

Myers, Norman. 1980. 'The Present Status and Future Prospects of Tropical Moist Forests'. *Environmental Conservation* 7 (2): 101–114.

Nordblad, Julia. 2021. 'On the Difference between Anthropocene and Climate Change'. *Critical Inquiry* 2021 (2).

Raup, David M., and J. John Sepkoski. 1982. 'Mass Extinctions in the Marine Fossil Record'. *Science* 215: 1501–1503.

Ripple, William J., Christopher Wolf, Thomas M. Newsome, Mauro Galetti, Mohammed Alamgir, Eileen Crist, Mahmoud I. Mahmoud, Willian F. Laurance et al. 2017. 'World Scientists' Warning to Humanity: A Second Notice'. *BioScience* 67 (12): 1026–1028.

Robin, Libby, and Will Steffen. 2007. 'History for the Anthropocene'. *History Compass* 5 (5): 1694–1719.

Rose, Deborah Bird, Thom van Dooren, and Matthew Chrulew. 2017. 'Introduction: Telling Extinction Stories'. In *Extinction Studies: Stories of Time, Death, and Generations*, edited by Deborah Bird Rose, Thom van Dooren, and Matthew Chrulew, 1–19. New York: Columbia University Press.

Secord, James A. 2004. 'Knowledge in Transit'. *Isis* 95 (4): 654–672.

Steffen, Will, Paul J. Crutzen, and John R. McNeill. 2007. 'The Anthropocene: Are Humans Now Overwhelming the Great Forces of Nature?' *Ambio* 36 (8): 614–621.

The New York Times. 2014. 'Books. Bestsellers. Hardcover Nonfiction'. *The New York Times,* March 30, 2014. https://www.nytimes.com/books/best-sellers/2014/03/30/hardcover-nonfiction/

The Pulitzer Prizes. 2015. 'The 2015 Pulitzer Prize Winner in General Nonfiction'. https://www.pulitzer.org/winners/elizabeth-kolbert, accessed June 19, 2019.

UN Environment. 2019. *Global Environmental Outlook, Key Messages.* https://www.unenvironment.org/resources/assessment/geo-6-key-messages, accessed October 21, 2019.

Warde, Paul, Libby Robin and Sverker Sörlin. 2018. *The Environment: A History of the Idea.* Baltimore, MD: Johns Hopkins University Press.

Wilcox, Bruce A. 1988. 'Tropical Deforestation and Extinction'. In *IUCN Red List of Threatened Animals,* edited by IUCN, v–x. Gland, Switzerland: IUCN.

Wilson, Edward O. 2001. *The Diversity of Life. New Edition.* London: Penguin Books.

Zalasiewicz, Jan., Colin N. Waters, Colin P. Summerhayes, Alexander P. Wolfed, Anthony D. Barnosky, Alejandro Cearreta, Paul Crutzen et al. 2017. 'The Working Group on the Anthropocene: Summary of Evidence and Interim Recommendations', *Anthropocene* 19: 55–60.

9 Smoke, smells, and seaweeds in eighteenth-century Norway

Anne Eriksen

Introduction

Travelling through Norway from the south towards the city of Trondheim in 1804, Christen Pram met peasants, fishermen, and civil servants, all of whom told him the same story: ash from the burning of kelp along the coast had changed the weather for the worse and was directly to blame for the succession of cold and rainy summers. The crops failed, and the fish were chased from the fjords by the heavy and foul-smelling smoke from the kelp kilns. This combination of failures left the coastal population starving. A ban on kelp burning was demanded (Pram 1964, 92). As a member and envoy of the Board of Trade (Commercekollegiet) in Copenhagen, Pram had been charged with the task of investigating these stories, which had been circulating for some time. Did the rumours hold true? Was kelp smoke really guilty of changing the weather and damaging grain crops and fisheries? The aim of the present chapter is not to answer these questions but to examine Pram's efforts to solve them. What kinds of information did he gather on his journey, and how did he work to process it? What kind of knowledge did he produce?

This discussion also relates to a more general issue. Pram set out to investigate what can be termed local natures: particular things that were going on in particular places, defined by 'the characteristic combinations of flora and fauna, climate and geology that give a landscape its physiognomy' (Daston 2019, 15). To achieve his task, however, Pram was to do more than just assess local conditions. Based on an understanding of nature as ruled by universal laws, Pram was going to apply tools and develop knowledge that was not restricted to the local. His investigation was to be built on knowledge about universal nature and the lawlike regularities that define 'a uniform and inviolable order, everywhere and always the same, exhibiting ironclad regularities' (Daston 2019, 23). In contrast to nature's local customs, this kind of general knowledge admits no exception, which implied that the knowledge resulting from Pram's work would be valid beyond the specific localities where it was produced. Even if the forms and regularities of nature had been studied since antiquity, 'the nature of natural laws – uniform, universal and inviolable – emerged in the course of the seventeenth century', Daston argues (Daston 2019, 25).

The separation of (local) nature from knowledge about natural laws has made it possible to act upon nature in transformative and decisive ways, to build the modern world. The story of how this separation emerged, or was produced, can be told in several different ways. This chapter does not seek to add to any pre-existing grand narrative about modernity and/or Western culture but will explore instead the process of separation on a micro level by means of a specific case. The aim is to examine not only how time or temporality represented important tools to achieve this separation but also how difficult this work could prove to be. Inscribing (universal) *knowledge about* nature in another temporal regime than the organic temporality of (local) *nature itself* contributed significantly to differentiate between them. Certain temporal frameworks made it possible to conceptualise knowledge about nature as non-local and general, not bound up with any specific time or place but belonging to the everywhere and always of science. Natures in the plural, on the other hand, could be left to specific localities and to be defined by the seasonal rhythm of the sea and the schooling fish, the land, the fields, and their crops. As a consequence, knowledge that sprang directly from these specific places and rhythms also came to be regarded as integral to the realm of nature rather than as part of knowledge about it.

It will be part of the argument in this chapter that the experiences, events, skills, and competences that were involved in 'the kelp affair' were based on different temporalities or ways of experiencing and organising time, both that of nature and that of knowledge about nature. In this way, efforts to understand what was happening to the fish, the crops, and the climate when kelp was burnt on the shores also contributed to the fundamental process of separating nature itself from new, scientific types of natural knowledge.

In modern parlance, the issues that Christen Pram was sent to investigate were 'climate change' and 'pollution'. In contemporary language, 'climate change' has become a term that effectively coordinates phenomena, times, and spaces, and makes it possible for us to organise experiences and events, to develop 'different kinds of knowledge and understandings, distribute commitments and responsibilities, facilitate different kinds of action, and outline possible futures' (page 5). Even if the climate and the possibility of man-made changes to it were much debated even in the eighteenth century (Bonneuil and Fressoz 2017; Eriksen 2019), 'climate change' was not a term equally convenient at hand in that period. It could not as easily be used to argue for the validity of or otherwise assess the complaints of the damaging effects of kelp smoke. People – Pram and the Board of Trade, as well as the people he met on his journey – would make use of other concepts and phrases in their efforts to 'distribute commitments and responsibilities, to facilitate different kinds of action, and to outline possible futures', in order to articulate what it was all about, and what ought to be done about kelp burning and the smoke it produced. The present chapter will investigate some of these efforts. What concepts and terms were employed to interpret and explain events and to carve out strategies and actions? What kinds of knowledge were invoked, and how were they used?

The empirical material for this exploration will be letters and notes written by Pram on his journey during late summer and early autumn 1804, as well as the report produced when he was back in Trondheim in October. Along the route Pram wrote letters to local contacts, mainly merchants and civil servants. Notes and memoranda were addressed to the Board of Trade, reporting on his travel and work and communicating the testimonies and opinions of the people that he met and talked to. The texts demonstrate that he actively sought out information from a wide range of different sources. They show that he met and conversed with local peasants and fishermen, as well as with civil servants and other local dignitaries along the route. He collected testimonials and evidence, rumours and observations. He also carried out his own experiments with fish, smoke, and water. In this way the material he gathered came to represent quite different types of knowledge. Popular tradition, the local fishermen's experience, and the ideas of local merchants and clergy blend with Pram's own knowledge of natural history and with his more or less systematic observations and experiments. These different types of knowledge, experience, and communication each came with their own type of authority. They also engage with time in different ways. Some of the knowledge is deeply embedded in local geography and natural conditions, as well as in traditional skills and customs along the coast. Other elements are external and universal, based on the authority of natural history and natural philosophy – what today we would call science. I will make use of Pram's notes and letters to trace his journey and his efforts to gather information *en route*. From this basis, I will proceed to explore how he made use of the different types of knowledge that he encountered, and how he processed this material in his report.

The kelp affair

The aim of the kelp burning was to produce sodium carbonate for glassworks and for the production of soap. The ash was partly exported to England and partly used by newly established domestic glassworks. The production represented a welcome source of income to the poorer part of the population along the north-western coast. It also answered to the authorities' wish to encourage new industries and develop the use of natural resources in the coastal regions. Nonetheless, this industry had been opposed ever since it started up in the 1760s (cf. Johannessen, 2020, for a thorough historical presentation; Locher and Fressoz 2012). The coastal population of fishermen and farmers had long demanded a ban on the activities, and in 1804 kelp burning was temporarily suspended. In June 1804, shortly before Pram set out on his journey, fishermen at the island of Smøla were reported to have attacked kelp workers and destroyed their produce. The authorities in Copenhagen acknowledged the need for more precise knowledge. The level of conflict in desperate local populations was high and it took little to make violence break out. A succession of summers with crop failures contributed significantly to this urgency. The need to detect possible reasons for the failures was pressing.

Kelp burning had been debated in newspapers and journals for some time. *Trondhjemske Tidende* – published in Trondheim, at the time Norway's second-largest city and close to the area of the kelp burning – gave attention to the issue in the late 1790s (cf. Withammer 1799a; 1799b). *The Royal Society of Sciences and Letters*, also situated in Trondheim, launched a prize competition on the matter and published the winning contribution (Rynning 1803). In his description of the city of Kristiansund, the local inspector of customs, F. W. Thue, discussed whether the smoke could be the cause of failing local fisheries (Thue 1796). The kelp affair also attracted attention in the capital of the twin-kingdoms, and the botanist and veterinarian Erik Nissen Viborg reported a number of experiments in Copenhagen to investigate the potential malignity of kelp smoke. He concluded that the smoke had a terrible smell and left a bad taste on fish, meat, and milk (Viborg 1805). Despite the urgency of the issue, none of these texts reflect attempts at empirical investigations of specific, local conditions. Viborg's experiments were strongly empirical but carried out in Copenhagen and with a species of kelp that did not grow in Norway. Withammer, Thue, and Rynning were all local, and they obviously knew the area and its geography rather well. Their arguments were nonetheless dominated by theoretical and rather general considerations about the nature of smoke, its interaction (if any) with water, and the behaviour patterns of herring and whales.

Reported events and published texts also show that the question of kelp burning and smoke tended to be entangled with a number of other issues. The earliest revolt, in 1765 (*Strilekrigen*), was above all about new taxes. Professor Johan Chr. Fabricius at the University of Kiel, travelling through Norway to collect information on natural history and economy in 1778, for his part reported that kelp burning and the smallpox inoculation were both regarded as a possible cause when the annual fisheries failed in the Kristiansund region. Popular opinion held that preventing smallpox by the 'artificial' means of inoculation (a kind of early vaccination) was an act of hubris that provoked divine wrath (Fabricius 1790). From Trondheim, both Withammer and Rynning wrote strongly in favour of the new industries for which the ash was produced, and obviously wanted to clear the kelp burning of the charges against it. They consequently suggested other reasons for the failing fish. Withammer pointed to overtaxing caused by new fishing equipment, while Rynning argued that natural changes in the schooling pattern of the herring was the most significant cause (Withammer 1799a; 1799b; Rynning 1803). The variety of perspectives and interpretations show that kelp burning was a different type of issue to different groups with different interests and social positions. What they nonetheless have in common is that all represent efforts to come to terms with kelp burning and its effects, to settle what it was all about: economy and the fair distribution of resources, new industries and enterprises, religion and divine will, or perhaps a mix of all this and more. In their very plurality, the versions of what was going on and why it mattered were attempts to 'distribute commitments and responsibilities, facilitate different kinds of action, and outline possible futures' (page 5).

When Pram arrived in Trondheim in 1804, his distinct aim was to see for himself what happened when kelp was burned. He wanted to observe the smell and smoke as directly as possible, in order to understand their eventual effects on fish and fishing conditions, weather, and crops. Only this empirical approach could produce the knowledge that was required. Pram emphatically refused to pronounce himself before proper investigations had been carried out, and declared that 'I will abstain from uttering anything in this matter until I, either by organizing a kelp burning or observing one in a place where it is reported to cause damage, can see for myself how the smoke settles and, if possible, experience how it works' (Pram 1964, 11).[1] Pram also had a clear plan for how to achieve his goals. Starting out from Trondheim, he would visit nearby Ørlandet, one of the areas where kelp burning had been going on before the temporary ban. Aided by the local vicar, Matthias B. Krogh, he would conduct an experimental burning when the herring came into the fjord to spawn. From Ørlandet, the plan was to continue southwards by sea to the cities of Kristiansund and Molde, visiting the larger islands on the way. En route, Pram would contact fellow civil servants who could supply him with information and speak to the locals. From Molde, he would then return to Trondheim to finalise his reports to the Board of Trade. The whole trip was intended to take three to four weeks, which implied that it would end in the beginning of August. But things did not go as planned.

Pram's journey

Pram arrived in Trondheim on 10 July. Four weeks later he was still there. By now he had written a lot of letters as well as memoranda to the Board in Copenhagen but was not much wiser when it came to kelp and smoke. There had been weeks of heavy rain. The herrings had not arrived in the fjords. The vicar Krogh, who was going to arrange the experimental kelp burning at Ørlandet, was away. The locals were still busy haymaking, and the fisheries had not yet started. Pram was a bad sailor, and the thought of travelling by sea along the coast in the bad weather did not tempt him much. His neatly planned timetable for systematic observation thus was overrun by other more natural temporalities: those of the weather, the migrations of the fish, and the seasonal work of the local peasants.

Similar complications followed him during his entire journey. Tensions between the temporality of nature as it was embedded in the cycles of seasons and the labour of harvesting natural resources, and that of knowledge production *about* nature, based on a pre-planned schedule and universal categories of time and space, came to saturate Pram's entire endeavour. The planned inspection, the systematic observation, and the rational experiment governed by a neat timetable were thwarted by the unruly forces and less-disciplined temporality of the natural conditions in the harsh coastal region. This temporality, intrinsic to the nature that was to be inspected and observed, was regular, but unpredictable in its specific, local manifestations. The travel plan, on the other hand, with its timetables and the dates with which Pram's notes and letters were neatly marked, reflect the cleansed, autonomous, and in itself empty time of the clock and the calendar, envisaged as a tool in the efforts to create systematic and non-local knowledge. From the very beginning of Pram's journey, this neat efficiency was upended by the natural force of winds, weather, and schooling fish.

On 9 August, Pram finally left Trondheim for Ørlandet. Neither the herring nor the vicar had shown up, but the weather had improved, and the proprietor of the Østeråt estate, Holtermann, had promised to help with the experiment. Due to the temporary ban on kelp burning, the experiment had to be conducted with care. The locals were still waiting for the herring schools and could easily be enraged by the sight of burning kelp. Pram also made use of his stay to collect information about the rural economy at Ørlandet before rumours of herring at the island Hitra then made him travel there. He arrived on 18 August and stayed for a week, visiting some of the smaller islands and gathering information about the fisheries. He also spoke to the merchant Parelius and the 'wise and experienced' vicar Brodtkorb, who introduced him to some of 'the most excellent and experienced fish farmers in this country' (Pram 1964, 78). He heard rumours about approaching herring and had the luck to observe at least one large seine catch at Hitra, as well as several smaller ones on the other islands.

Pram then continued southwards by land via Vinje at the bottom of the fjord, and from there by sea to the city of Kristiansund. Due to a strong headwind, the last passage of no more than 25 kilometres took the whole of twelve hours by rowboat. On arrival, he was met by rumours of another kelp uprising at the island of Smøla. Speaking with the local chaplain Bull, who was also responsible for the remote island of Grip, Pram was informed about the precarious living conditions on the outer islands. He learned that the fishermen there had recently lost three days of work to a heavy kelp smoke which prevented them from navigating, which would severely affect their supply of food (Pram 1964, 80). The weather turned bad again, with freezing cold temperatures and the sea so harsh that even the post-boat was shipwrecked. Pram was taken ill and had to remain in Kristiansund until the end of the month. On 8 September, he finally arrived in the neighbouring city of Molde.

The heavy rain and strong winds continued. Pram was still quite unwell, but after a week he received an important visit. For the sole purpose of seeing Pram, a group of fishermen had travelled to Molde from the small outer island Ona; they had come to 'ask for [his] cooperation so that the kelp burning which is so devastating to the fisheries be totally banned, at least for some years' (Pram 1964, 84).[2]

The men are emphatically described as *honourable*, *old*, and *experienced*. A week later, Pram was on his way back north. Not only did travel in bad weather again take much time, but his progress was also delayed by social obligations. Pram took part in the wedding of a daughter of the Dean Budde at Bekken. Not until early October was he back in Trondheim, nearly three months after his first arrival there. Instead of the planned three weeks, the inspection tour along the coast had taken the larger part of two months. Pram could finally sit down to write his report. He summed up the main events of his travel and sought to make use of this information to answer the main questions of the case: Does the burning of kelp damage the fisheries? If so, how? Does the smoke really change the weather and bring damage both to the fisheries and to the crops?

The report shows that, since he set out to execute his experiment, Pram had learned a great deal about the weather, the land- and seascape, and the precarious living conditions along the coast. He had met with storms and heavy rain, and he had spoken to people whose entire existence depended on hard work onboard small boats in the open sea. For himself, he had experienced seasickness, bad colds, and the fear of shipwreck in an open boat in a strong headwind. On his return to Trondheim, he knew far more than before about being subjected to the brute force of nature and about being governed by this temporality rather than by that of a pre-planned, written schedule. These experiences shine through in the report he wrote and in his (inconclusive) answers to the initial questions about kelp burning and its smoke.

Pram had started out on his journey as a detached observer, determined to build on nothing but his own experiments and systematic observations, but he returned with funds of embodied knowledge and corporeal experiences, all deeply embedded in his encounters with the people and nature along the coast. These entanglements stand out as a common and increasingly important theme in all his writings relating to the journey. In the final report, they become a conspicuous reflection of Pram's struggle to order his material and extract exact and systematic knowledge.

Knowledge and authority

The report reflects Pram's efforts to deal with the experiences described above and to convert them into knowledge that would help him answer the issues that he had set out to investigate. However, the different temporal regimes inherent to his experiences, on the one hand, and those stemming from his plans and intended method, on the other hand, produced twists and turns in the information and knowledge that he sought to extract from those experiences. The temporality of nature and that of scientific knowledge each structured events and created meanings according to their own different schemes. Pram worked hard to keep them separate, but not only did the time of nature and that of scientific knowledge keep entangling, they also marked his own experiences.

Did kelp smoke affect the weather for the worse? This is the introductory topic in Pram's report. On his travels, Pram had met people who expressed their deep gratitude to the government for the temporary ban on kelp burning. Thanks to this, they declared, the country had regained the good summer weather that had been sorely missed in recent years. Everybody now hoped for a good harvest. The burning of kelp was considered to be, if not the only, undeniably a subsidiary cause of the sky's constant cloud cover during recent summers. In consequence of the continuous bad weather, the grain had not ripened and the small amount that could be harvested was rotten and musty, people said. The present year was different, with very good weather during spring. In mid-June, things had changed, however, and three weeks of rain followed. Nobody doubted the reason: the production of kelp ash had been resumed (Pram 1964, 92). In Trondheim, Pram had heard rumours that when kelp burning was started at the island Smøla, the air immediately turned grey and rain started to fall. The local population had gathered, attacked the kilns, put out the fire, and thrown the amassed kelp into the sea. The results were not long in coming: 'The sky nearly immediately became serene, the air warm and fertile, and everything promised a good harvest' (Pram 1964, 93).[3] Nobody asked for further proof: kelp smoke changed the weather.

Pram does not comment on these rumours and claims until the final section of the report. In this way, he makes the issue of weather and climate frame the entire text. When he returns to it, his phrases are somewhat more cautious than in the introduction:

> The fact that is reported is that because kelp in some quantities is burnt annually, the climate in these areas or this part of the country has become rainier and colder than before, when little or no kelp was burnt.
>
> (Pram 1964, 110)[4]

Whether components of the drifting smoke can turn the ever-humid air into showers is for 'better physicists than me to investigate',[5] Pram declares (Pram 1964, 110). On his journey, however, Pram had experienced the local sea breeze called 'havgule', which regularly rose during warm summer days. Coming in from the open sea and following the fjords, this wind was cold and humid. Could it cause clouds to gather and rain to fall when it encountered the kelp smoke?

> Do the components of the kelp smoke have some special affinity to heat, by which it is being extracted by the arriving cold and humid air, and by this lets the vapours gather into fog, clouds, and rain; do they perhaps effect the said phenomenon according to their qualities and the nature of the matter in some other way; or is the whole thing just a fancy?
>
> (Pram 1964, 110)[6]

Pram's discussion of these questions represents a mixture of general statements about the sea breeze and more specific testimonies concerning the allegedly damaging effects of kelp smoke on crops, meadows, and fish during recent years. Even if it is certain that the last summers, when kelp had been burnt, had had fairly bad weather, nobody knows how it would have turned out without the smoke. And 'neither [does] the physicist', Pram ends his reasoning (Pram 1964, 111).

Pram's argument reflects the way that he negotiates different types of knowledge in his work to find answers to the questions about the possibly harmful effect of the smoke. His own observations, the testimonies of local people – 'the most intelligent and reliable men in the parish' – and the knowledge represented by the invoked physicist, are counterbalanced and considered. These types of knowledge also represent different types of authority. Some of them are internal and informal, based on the experience, the social reputation, and personal 'honour' of the informants. In Daston's terms, this knowledge largely refers to 'local nature'. Other types of knowledge that Pram evokes are external, general, and systematic, and refer to 'universal nature'. This knowledge is represented in the text by the references to physics and to the absent and hypothetical physicist. Pram works to translate between the two types of knowledge as well as between the two notions of nature that they imply. He acknowledges both the external and the internal knowledge, but doing so does not help him to arrive at definite answers.

This discussion reflects the ambivalence of the entire report. Pram negotiates physical laws and other scientific knowledge about natural regularities with experience – his own as well as that of the local population. The aim is to use general forms of knowledge to interpret and explain local specificities, and in turn to produce insights that can be lifted out of their immediate and empirical contexts. However, the result most often is inconclusive.

The experiment

The experimental kelp burning is given pride of place in the report. The first third of the text is devoted to a description of what was done and of the knowledge that was inferred. In this way, the experiment, which actually was completed in a couple of days, appears as a core element of Pram's work and is set up as a key to understanding the kelp issue. The presentation is full of exact measures of time, size, and volume. The kiln that was built for the purpose of the experiment was two feet wide, one foot deep, and fifteen feet long. The pile of kelp was about two feet high when the kiln was lit. This was done at half past three in the afternoon. The kiln was lit on one end and, in the course of one hour, the fire had reached the opposite end. The kiln produced heavy smoke, but as the wind changed three to four degrees every half hour during the afternoon, the clouds of smoke were regularly dispersed. In the late afternoon, the wind shifted westward, which left the smoke to rest on the water of the fjord, hiding everything along the southwestern coast for a distance of two miles. In this position, it remained for two to three hours before dispersing. In the evening, the wind shifted even further to the west and filled the entire fjord with smoke, making it extremely difficult to discern any objects at all. The kiln was refilled as the kelp burnt out, but after five and a half hours, at nine o'clock in the evening, the entire mass of kelp had been consumed. Pram calculates this to have been 495 cubic feet (Pram 1964, 94–96).

In order not to provoke the peasants, the experiment was executed with great care and the kiln consequently had been smaller than the ones in regular use. In his report, Pram systematically compares the measures cited above with those of a regular burning in a kiln of ordinary size. The amounts of kelp and smoke were correspondingly larger, and the entire process likewise took more time. According to Pram's calculation, a regular burning would take two to three days and consume about 22,500 cubic feet of kelp, with the kiln being refilled once an hour. The kiln usually was 50 feet long and 3 feet wide, while the height of the regular kiln was identical with that used by Pram (Pram 1964, 96).

The strong emphasis on exact measures makes the reduction in size and duration of the experimental burning appear as a systematic down-scaling, producing a neat miniature. Even if Pram regretted the reduction, caused by practical constraints, it contributed to turning the experiment into a model. By means of limited scale and exact measurements, the rather messy affair of setting fire to a pile of wet kelp and feeding the smoking, reeking, and smouldering heap for days, was turned into a controlled and well-proportioned experiment. Substituting the temporality of nature with that of the clock is a core element of this process. The knowledge that could be gained from the burning was still situated and local, but as a model the event also represented a repository of knowledge to be lifted out of its specific context. The local event was transformed into autonomous and potentially general knowledge.

Pram's presentation of the experiment is accompanied by descriptions of how the smoke behaved and looked, how it smelled and tasted. Pram continued his observations of the smoke into the dark of the night, noting that it was reported to have dispersed at five o'clock in the morning. He notes with surprise that kelp smoke did not much impede breathing, even when one stood in the middle of the cloud. Not even Holtermann, the proprietor of the Østeråt estate who assisted at the experiment and who suffered from a lung disease, was much troubled by it. According to the local surgeon in Molde, however, it could be harmful to eyesight.

Pram also reports that the smell of burning kelp reminded him of burning wool or feathers, or a mixture of both. The smell clung to clothing so strongly that for a long time he was not able to touch the garments that he had worn during the experiment. Merchants in Kristiansund had told him that even among the most ill-smelling fishermen, people engaged in kelp burning can be discerned due to the smell (Pram 1964, 99). As for taste, Pram had arranged for a pail of water to be standing close to the kiln and the smoke, expecting the water to thereby acquire a foul taste. However, he did not notice any particular taste to the water, which surprised him, and which was, he writes, contrary to the experience of others. He refers to anecdotes of people with different experiences. Perhaps the position of the pail was the reason, Pram considers, and goes on to explain that it was standing very close to the kiln, which meant that the smoke was still very warm when it passed. For this reason, it might not have penetrated the water but merely passed over its surface without leaving traces (Pram 1964, 98).

In this part of the text, Pram seeks to make use of his experiment to infer further knowledge. He tries to articulate insights about smoke, taste, and other effects of the smoke in terms of generalised knowledge. He is not too successful. One main reason for his failure is that the results from the experiment were not very decisive, and consequently there was not much he could speak with any certainty about. Still, the argument is a continuation of the process that started when the burning was transformed from a local event to an object of autonomous, non-local knowledge. The apparent exactitude created by framing the events in the terms of exact measurements and times served to cut them off from their specific contexts and lay them open to further generalising.

It remains a major problem, nonetheless, that the experiment was not executed as originally planned. This does not refer to scale but to fish. After all, Pram's main concern was not the smell and taste of the smoke as such, or the ways humans might react to it, but the behaviour of the fish. Did they shy away from the smoke? Pram had planned to have the experimental kiln lit when the seasonal schools of herring arrived in the fjord. This plan was why he stayed so long in Trondheim – the migrating herring had not arrived at its usual time. When he finally gave up waiting and started the experiment at Ørlandet, it was mainly because further delay would have postponed the continuation of his further journey too far into the bad season. Consequently, the experiment that had been designed to find out whether the fish really shied away from the kelp smoke was actually carried out without schooling fish in the vicinity.

Another concern was whether fish were harmed or died from kelp smoke when they could not escape it. To find an answer, Pram arranged for a catch of small fish of different species to be placed in two large tubs that were both filled with water,

> of which the one should be placed where no smoke could reach it, and the other in a building where a large mass of kelp would be lit, to observe whether this would kill them. The fish were arranged in this way in the evening, but before the experiment could be carried out, the fish in both tubs were dead.
>
> (Pram 1964, 97)[7]

In this situation, Pram had to content himself and the readers of the report with describing his own reactions to the smell, taste, and look of kelp smoke.

He adds that he did not find it advisable to arrange any further experiments and gives two reasons for the decision. The first is that every person he had spoken to was firmly convinced about the damage caused by the smoke and considered the evidence for this to be overwhelming. The other is that the local population at this time was eagerly waiting for the seasonal herring, reported to be seen along distant coasts and expected to arrive in the fjord soon. Tools and equipment were ready and people were gathering and preparing to take to the sea once the signal was given. Pram concludes that

> [w]hen the herring already had arrived in a fjord, it would be wrong to venture to chase it away by a burning, and thus forfeit the catch of several hundreds, even perhaps thousands, of barrels of herring, and where no fish was, a repeated or larger burning would not produce more information than what had already been gained.
>
> (Pram 1964, 100)[8]

This conclusion marks the end of the first section of Pram's report. The rest of the report is not written on the basis of more experiments but instead on what Pram calls 'the assertions and testimonies of experienced men' together with 'the nature of the matter' (Pram 1964: 100).[9]

Knowledge and experience

This shift in Pram's argument illustrates how the experiment and the scientific knowledge that it was intended to produce was gradually eclipsed by the temporalities of the natural phenomena that this knowledge was supposed to be about. Most conspicuously, this happened when the planned focus of the investigation – the schooling fish – was absent from the fjord during the single experiment that actually was conducted. The exact time and measurements of scientific knowledge did not prove able to discipline the seasonal migrations of the fish. The catch of fish that unexpectedly (but probably quite naturally) died in the tubs prior to the scheduled experiment is another example. Neither the life cycles nor the seasonal cycles in nature would adhere to the abstract temporality of knowledge about nature.

The announced shift in the text can be understood as an attempt to come to terms with this challenge by shifting to another strategy. The temporality of knowledge that Pram sought to impose by means of the exact time, scale, and measurements of his experiment extends to a more general scientific 'longue durée' consisting of natural laws and regularities – referred to by Pram with the term 'the nature of the matter'. To this is added the correspondingly long-term knowledge of 'experienced men'. The temporality that marks this type of knowledge is frequently presented as an opposition between 'before' or 'as it used to be' and 'now' or 'recent summers'. Once again, two different temporalities – of nature and of knowledge about nature, respectively – are put into play. Both are general and long-term, but the one is still embedded in local experience, geography and custom, while the other is universal and abstract.

Pram accepts as true the numerous stories about how the fish have disappeared from the fjords, or at least that the size of the catches has been significantly reduced, and tries to find the reasons for this change. His argument is based on the rather inconclusive results from his experiment together with more general statements about the behaviour of fish, smoke, and smell. It seems improbable, he writes,

> [T]hat the fish, even if the surface of the water was poisoned by the smoke, for this reason should shy completely away from the fishing banks and grounds, as the effect of smoke hardly could reach deeper than the upper parts of the water [...] But it is partly uncertain if the pollution of the water, even if it is imperceptible to our sensory organs, nonetheless could be strong enough to drive the fish into the depth of the waters [...].
> (Pram 1964, 103)[10]

Moreover, it is unknown exactly how the fish will be affected when it emerges from the depths to catch insects on the surface of the water. The instinct of the fish is to follow the insects on which they feed. It is also to flee the whales and dolphins that chase them, which is the reason why large schools of fish appear in the fjords. All these creatures live on the surface of the water or need to come up to breathe. So perhaps the fish have disappeared as a consequence of the changed behaviour of these other species?

To these deliberations based on a general and non-local knowledge about 'the nature of the matter', Pram adds statements about local nature and its changes over time. The opinion is unanimous, he writes, that the fisheries were in greater bloom 'in earlier times'. He gives figures and numbers as evidence, informing his readers about the size of herring catches in earlier years and the number of barrels of cod liver oil that was previously produced (Pram 1974, 109). During his travels, he had also learned that the fish stocks in many places along the coast were significantly reduced compared to earlier centuries when the fisheries had supplied a living for several hundred families. One of these villages even had had its own church (Pram 1964, 74).

The considerations do not lead to any exact answer. Pram proceeds to an evaluation of the testimonies he has collected but ends up with a discussion about the conflicting interests of his informants. He notes that persons who gain from the kelp burning hold it to be quite harmless. They have also largely succeeded in establishing as true that the opposite opinion is 'superstitious' or prejudiced. People engaged in the fisheries, on the other hand, condemn the smoke as detrimental even if some of the poorer fishermen dare not oppose the elite by expressing their own 'superstition' (Pram 1964, 101). In this way, 'experience' gives no more exact knowledge than 'the nature of the matter'. The investigation remains inconclusive.

The report includes the momentous experience in Molde when Pram was sought out by a group of fishermen from the island Ona, 'with the honourable, very ancient man Ole Knudsen as its leader'. Like other locals they described how all fish shied away from the kelp smoke and that when no kelp was burnt the fisheries were as rich as ever. But most important of all, the men could inform Pram about an experiment carried out some weeks before close to their own home. A large amount of herring, saithe, and cod had been closed off in a seine while kelp smoke was led over it, 'upon which all the enclosed fish ascended dead to the surface of the water' (Pram 1964:102).[11] This was exactly the experiment that Pram himself had wanted to do! The results seemed conclusive, moreover, but other aspects probably were equally important.

In this experiment, as it was reported to Pram, the two different temporalities that he was struggling with in his argument actually met and merged. On the one hand, the experiment represented the same kind of detached knowledge about nature as his own work had tried to achieve. On the other hand, it was not only more successful but had also been carried out by men who literally embodied the temporality of nature itself *as well as* the kind of knowledge gained from long-term experience. The leader of the group, who is emphatically presented as both honourable and extremely old, becomes the very token of this amalgamated knowledge. Pram was so enthused by what the men told him that he immediately wrote to the local authorities at Ona to have the story confirmed (Pram 1964, 102). The fact that it was later discovered to be a hoax or at best a rumour is not included in Pram's report. That information can be found as a note in the latter part of his writings.

Despite this moment of excitement, Pram's discussion ends on a more resigned note. It simply cannot be proved with any certainty, he writes, whether fish die or shy away from the smoke. The inverse remains similarly unproven – that the smoke does no harm. But despite this lasting uncertainty, 'one is not justified in denying the truth of the testimonies nor the reality of the matter, even if the exact way in which the smoke causes its effects cannot be clearly realised' (Pram 1964, 102).[12] In this conclusion, the detached investigator, who set out to gain knowledge about nature by means of exact time, measurements, and the precise observations these tools would supply him with, has been transformed into a defender of the testimonies of experienced men and of knowledge that springs from close contact with nature's own cycles, rhythms, and seasonal changes. Even if he cannot present a definite answer to the issue he was commissioned to investigate, he has gained a new epistemological certainty, based on his own experiences with an experiment, observations, and encounters with living people in a harsh nature. Embodied and embedded knowledge, governed by the seasonal rhythm of the sea, the land, the fish, and the crops, and tied to specific places where hard-working people strove to gain a living, came to appear far more relevant and real than the general and non-local insights generated by science, systems, and the abstract time of the clock, the calendar, and detached observations.

Conclusion

The term 'climate change' was not available to Pram as the convenient shorthand term that it is today. It could not help him to sort and analyse the experiences and observations that he made. It is nonetheless obvious from the investigations above that the major problem Pram had to tackle was not a lack of adequate and functional terminology but rather the wealth and complexity of information that he encountered on his journey. From this perspective, there are obvious and significant parallels between Pram's work and our contemporary efforts to speak about climate change. These parallels concern in part the issue of scaling, and in part the very different types of knowledge that are not only involved but deeply entangled.

Pram sought to make temporality a tool for transposing knowledge about local nature into the domain of universal natural laws. By means of experiment and rigid observation, the temporality of universal natural laws was to be imposed on particular, local conditions. These laws, in their turn, were supposed to supply reliable and non-local knowledge about what was going on locally and about what ought to be done. The efforts represented a rescaling from local particulars to universal regularities and then back to local regulation. Reliable knowledge was to be gained, but it is equally important that this movement also turned the universal scale into a source of normative authority. When the knowledge produced by this method was transposed back onto the local and particular, it would also come with instructions about what should be done. In Pram's specific case, it would dictate whether kelp burning should continue or be banned.

What Pram experienced during his journey was the gradual breakdown of this normative hierarchy. Not only did the experiment and the observations prove impossible to carry out in the precise and orderly way he had planned, thus destroying the vital channel of transportation between the local and the universal; the mass and complexity of local, particular information reduced the relevance of the universal and lawful. The precariousness of living conditions, the harsh natural environment, and the fishing population's constant work to survive got the better of Pram's official knowledge production. Local knowledge, based on experience, embedded in local nature and an integral part of life there, marginalised the impact of systematic and scientific insights. The real challenge for Pram, then, proved to be not the execution of his planned experiments but the assessment of the overwhelming mass of this other knowledge. To be made useful, it could *not* be separated from its local context, it could *not* be cleansed from social and economic circumstance, and it could *not* be rescaled as universal science.

Notes

1 Jeg afholder mig fra at yttre noget angaaende denne Sag, saalænge til jeg ved at anstille en Tangbrænding eller paa flere Steder, hvor den siges at være skadelig, selv faaer see hvorledes Røgen lægger sig, og om mueligt, faar anstillet Erfaring om hvorledes den virker.

2 [...] anholde om min Medvirkning til at Tangaskebrændingen som ødelæggende paa Fiskeriene aldeles maatte forbydes, i det ringeste paa nogle Aar.

3 [...] at Himlen næsten umiddelbar derpaa blev klar reen, Luften Varm og grødefuld, og alting lovede en riig Høst.

4 Factum som man angiver, er at siden der aarligen brændes Tang i nogen Mængde, har Veirliget i disse Egne eller i denne Deel af Landet været meere regnfuldt og koldt end forhen, da der lidet eller intet brændtes Tang.

5 [...] Gienstand for Undersøgelse af bedre Physikere end jeg er...

6 Have nu Tangrøgens Bestandeele nogen særdeles Affinitet til Varme, hvorved den uddrages af den indstrømmende fugtige Luft, og giør derved at Vanddunsterne samle sig til Taage, Skyer og Regn; virke de det paaankede Phænomen maaskee i Følge deres Beskaffenhed og Tingenes Natur paa nogen anden Maade; eller er det helle blot Indbildning.

7 [...] af hvilke den eene skulde staae paa et Sted hvor ingen Tangrøg kom hen, den anden i et Huus hvor man vilde have antændt en heel Deel Tang, for at see, om disse derved skulde dræbes. – Fiskene hensattes saaledes om Aftenen; men inden man dermed kunde anstille Forsøget, var Fisken i begge Ballier døde.

8 Naar Silden allerede var indkommen i en Fiord, var det Uret at vove at jage den bort ved en Brænding, og derved maaskee forspilde Fangsten af flere Hundrede, ja maaskee tusinde Tønder Sild, og hvor ingen var, var der heller ikke ved gjentagen eller større Brænding meer Oplysning at hente end den der nu var hentet.

9 [...] deels efter Sagens Natur er at bedømme, deels nærmere ved Erfarnes Udsagn og Vidnesbyrd at oplyse.

10 [...] at Fisken, om end Vandets Overflade ble forgivtet af Røgen, skulde for dens Skyld flygte ganske bort fra hine Grunde og Stader, da dog Røgen synes umueligen at kunne strække sin virkning længer end til den øvre Vandskorpe [...] Men, deels er det uvist, om dog ej Vandets Besmittelse, om den end var ukiendelig for vore Sandseredskaber, dog indtil noget betydelig Dyb kunde være stærk nok til at fordrive Fisken ...

11 [...] et Forsøg med at indeslutte i en Nod en Mængde Sild, Sej og Torsk, medens man røgede over Nodten med Tang, hvorpaa alle de i Nodten indsluttede Fiske skal have svømmet døde op.

12 [...] saa er man uberettiget til at nægte disse Udsagns Sandhed eller Sagens Virkelighed, om man end ej klarligen kan indsee paa hvad Maade Røgen virker dette Phænomen.

References

Bonneuil, Christophe, and Jean-Baptiste Fressoz. 2017. *The Shock of the Anthropocene.* London: Verso.

Daston, Lorraine. 2019. *Against Nature.* Cambridge, MA: MIT Press.

Eriksen, Anne. 2019. History, Exemplarity and Improvements. Eighteenth Century Ideas about Man-Made Climate Change, *Culture Unbound, Journal of Current Cultural Research* 11 (3–4): 353–368.

Fabricius, J. Chr. 1790. *Professor I.C. Fabricius's Reise igjennem Norge i Aaret 1778. Med Anmerkninger over Naturhistorie og Oeconomie.* København.

Johannessen, Finn Erhard. 2020. 'Tangaskebrenning – en miljøtrussel for over to hundre år siden?' *Historisk Tidsskrift* 99 (3): 197–211.

Locher, Fabien, and Jean-Baptiste Fressoz. 2012. 'Modernity's Frail Climate: A Climate History of Environmental Reflexivity'. *Critical Inquiry* 38 (3): 579–598.

Pram, Christen. 1964. *Kopibøker fra reiser i Norge 1804–06.* Oslo: Norske kunst- og kulturhistoriske museer.

Rynning, J. 1803. *Tanker om Tangbrændingens Indflydelse paa Fiskerierne og Agerdørkningen*. Trondheim: Stephanson.

Thue, F. W. 1796. 'Beskrivelse over Christiansund'. *Topographisk Journal for Norge* 4: 94–96.

Viborg, Erik Nissen. 1805. 'Beretning om de Forsög, som det kongl. danske videnskabers selskab har ladet anstille for at undersöge hvorvidt at Tangrögen kunde være skadelig for Fiskene i Havet og Vegetationen'. *Det kongelige danske Videnskabers-selskabsSkrivter*, 206–230. København: S.Popp.

Withammer, Olaus. 1799a. *Patriotiske Tanker fremsatte ved Olaus Withammer Nidarosiensis i Anledning af Tangbrænderi-Feiden*. Trondhiem: Stephanson.

Withammer, Olaus. 1799b. *Fortsættelse eller Tillæg til mine patriotiske Tanker om Tangbrænderiet og Fiskerierne i Trondiems Stift*. Trondhiem: Stephanson.

10 Origin myths from the cultural historical archive of the Anthropocene

Vico, Burnet and the time of the deluge[1]

John Ødemark

Introduction: Reconciling Vico's divide

In the seminal article 'The Climate of History: Four Theses', historian Dipesh Chakrabarty observes that the Anthropocene calls for a new compact between the genres of natural and cultural history, and their vastly divergent timescales (Chakrabarty 2009). A new historiography, able to cope with the pressing challenges of the Anthropocene, must bridge the gap between natural and cultural history. This is because the reality of the Anthropocene has itself already erased the nature-culture divide; there can be no separation of natural and cultural history if humanity is the 'cultural' cause of global warming and species extinction frequently associated with the Anthropocene as the 'human epoch' (Chakrabarty 2009).

In what he calls a 'thumbnail sketch' of early modern developments in historiography, Chakrabarty traces the nature-culture divide back to the Neapolitan philosopher Giambattista Vico (1668–1744), who published his new science of history in three versions from 1725 to 1744. Vico's division, and the concomitant disregard of natural history, 'was to become a part of the historian's common sense in the nineteenth and twentieth centuries', Chakrabarty adds (2009, 201). He cites Vico's use of the so-called verum-factum principle as the epistemic formula behind the nature-culture divide in historiography. In Chakrabarty's wording, this epistemological principle comprised the idea that we, humans, could have proper knowledge of only civil and political institution because we made them, while nature remains God's work and ultimately inscrutable to man (Chakrabarty 2009, 201). Apparently, 'nature' is here excluded from a historiography solely concerned with human creations (like 'civil and political institutions') – what we, the intellectual heirs of Vico's distinction, would call *culture*. 'Vico scholars have sometimes protested that Vico did not make such a drastic separation but even they admit that such a reading is widespread', Chakrabarty contends (Chakrabarty 2009, 202).

My aim in this chapter is not primarily to protest Chakrabarty's historiographical origin myth but rather to fill out his thumbnail sketch with more textual and historical detail, focussing upon the complexity of the divisions and contractions between natural and cultural history. My inquiry will show that Vico reworks evidence from the *historia naturalis, historia sacra* and the *historia civilis*. In doing this, he aims to separate the Christian logos from fables and myths. This work takes the form of a complex art of boundary maintenance between natural and cultural kinds of history. Thus, traffic between various genres of natural and cultural history persists, even in the supposed author of the modern distinction in the philosophy of historiography. If, as Chakrabarty asserts, the Anthropocene calls for a new compact between natural and cultural history, a detailed account of how 'we got there' in the past can perhaps also point forward to a new mode of convergence between these genres.[2]

Modernity theory and Whig historiography have customarily identified the emergence of modern schemes, such as the nature-culture divide, in an early modernity defined as a harbinger of mature modernity. Historian of science Lorraine Daston has pointed out the persistence of an early modern origin myth telling the – paradoxical – tale of the simultaneous rise *and* fall of nature. If nature falls from semantic grace, loses its soul by becoming a disenchanted domain of meaningless causes between the sixteenth and eighteenth century, it will also ascend as the objective arbiter of all and everything, i.e. a new kind of authority with scientific but also social application (Daston 1998). 'Even those who', Daston adds, 'like Bruno Latour, loudly assert "we have never been modern" take this seventeenth-century moment to be seminal of our characteristic brand of unmodernity' (Daston 1998, 149).

In contrast, salient trends in cultural history and the history and philosophy of science have questioned such tales of origin and construed the early modern past as a '*foreign* country', and thus (at least metaphorically) irrelevant to the present knowledge landscape. This has often been accomplished by applying some version of a concept of bounded cultures that construed past periods as instances of different 'epistemes', 'paradigms', or 'styles of reasoning'. The historiographical commitment to the cultural difference of the past (encapsulated by the slogan 'the past is a foreign country') requires the identification of an object *in the past* that differs from the present in a manner analogous to how contemporary cultures differ spatially. Cultural historian Peter Burke has described such a commitment to difference in terms of translation:

> Consider the following recurrent problems in cultural history [...]. How is it possible to be able to translate every word in a text from an alien (or even half-alien) culture, yet to have difficulty in understanding the text? Because – so one is able to say if one adopts this approach to the past – there is a difference in mentality, in other words different assumptions, different perceptions, and a different 'logic' – at least in the philosophically loose sense of different criteria for justifying assertions – reason, authority, experience and so on.
>
> (Burke 1997, 165)

It is in cases of such mistranslation, Burke asserts, that we have to account for cultural difference (Ødemark and Engebretsen 2018).

My aim, then, is to fill out Chakrabarty's 'thumbnail sketch' *qua* origin myth, by shifting to another historiographical *scale* – to a textual and intertextual microhistory rather than the grand scale of modernity theory and its myths of new, cosmological demarcations. I will assume that Vico is not (only) the founder of a new paradigm in which we still live at our peril, and from which there is no escape, but that he (also) represents a certain *cultural otherness*, in Burke's phrasing, 'different criteria for justifying assertions – reason, authority, experience and so on'. This, moreover, involves construing him as a part of a domain he himself 'invented': namely, culture as collective human difference (Ødemark 2011). Consequently, I will also re-examine the epochal divide attributed to Vico attentive to what I will call here the *textual microhistory* behind the distinction. This implies being attentive to the following:

1 Vico's actor concepts and language games in the manner of cultural and intellectual historians such as Peter Burke (2007), Quentin Skinner (2002) and G. E. R. Lloyd (1990; 2004) – and what I, referencing the philosopher of science Isabelle Stengers, will call his '*matter of concern*' (Stengers 2011). With this last term, the *explanandum* of the *Scienza nuova*, the phenomenon Vico seeks to explain, is also related to a cultural and existential concern beyond 'mere' scientific problem-solving.
2 The textual and poetic organisation of the arguments, the text-building (Becker 1995) and the 'workness' of the texts under construction (LaCapra 1984), both across Vico's texts, following the textual history of some of his key concerns, and in the section of the *Scienza nuova* of 1725 where he deals with Thomas Burnet.

I shall follow two convergent lines of inquiry. First, I will examine the notions of fable, fabulous history and periodisation in Vico. Now, 'fable' and its cognate 'myth' share an ambivalent semantic space; both terms refer to concrete narratives as well as superstition and epistemological errors – the opposite of the religious or scientific logos.[3] Second, I will close-read a section of the *First New Science* (1725) where Vico briefly tackles Thomas Burnet's *Telluris Theoria Sacra*. Burnet was an English theologian who investigated what he saw as the sacred history of the Earth, a theocentric project that has made him a villain in the history of geology (Gould 1987, 4). Both Vico and Burnet set out to defend Christian chronology against the 'deep time' chronologies explored in what we would call natural (emergent geology) and cultural history (accounts of Chinese and Egyptian chronologies). Hence, guarding sacred history and Biblical chronology against what historian of science Paolo Rossi called the 'dark abyss of time' is a matter of concern in both emergent geology and cultural history (Rossi 1984).

Several commentators have linked the textual meeting between Burnet and Vico to the Neapolitan philosopher's disdain for natural history. Rossi cites the Vico scholar Pietro Piovani on how Vico developed a 'philosophy *without nature*' and a 'philosophy of culture [...] as the heir of a defunct philosophy of nature' (Rossi 1984, 105). But what genres of natural history did Vico distance himself from? Was there one clear break with one kind of natural history, or a variety of relations between natural and cultural forms of history?

A fable within a fable: Vico's origin myth and Chakrabarty's

There is a certain recursiveness in using Vico and his distinction to mark the entrance to an epoch in the history of historiography governed by the separation of natural from cultural history. Vico himself was obsessed with origin myths and historical beginnings, and it is a conventional gesture in the history of the human sciences to plot Vico as a 'beginning' (Said 1997).[4] Vico himself, moreover, was deeply concerned with how historians should account for the first periods of human history, what he – in the tradition of Roman historian Varro – called 'obscure' and 'fabulous' history (e.g. Rossi 1984, 158). Vico was adamant that the first task of his new science was finding a method that actually could explain fables, and thus harness them as sources to a 'pre-history' that lacked written sources (Vico 2000, 329).

Modern historiography and its philosophy are passionately concerned with temporality. This fascination conceals the fact that (apparently) temporal categories such as periods and epochs are also spatial enclosures, where the flow of time is regulated by external boundaries. Within these boundaries, certain manners of thinking, acting, and narrating are supposed to be paradigmatic, i.e. both 'govern' *and* be 'instances' of the forms of thinking, narrating and acting considered possible in a particular historical enclosure or period.

Moreover, periodisation is consequently inherently *chronotopic*: it is dependent upon a *spatial* framework that furnishes the narrative *background* against which exemplary tales can illustrate what might be typical of a period, age or culture (Bakhtin 1981, 198; Puckett 2016, 157). Within this spatial enclosure objects can be 'timed' or given historical 'value' as a part of a paradigm or more comprehensive pattern that can be exemplified by concrete instances such as Vico's division – and tales about its more or less toxic afterlife as the 'rule' of modernity. Regularly these new rules are represented in macro-historical tales about modernity and secularisation as a new 'constitution' that separates nature from culture and restricts the action sphere of God to the heart of the individual. A case in point is the doyen of science and technology studies, Bruno Latour. For Latour, the modern is based upon not only a *separation* of nature and culture but also the continuous processes of *translation* or *mediations* that link nature with society, which in turn are balanced by processes of purification that restore the divide (Latour 1994; Bauman and Briggs 2003).

A similar logic of periodisation is worked out in relation to the narrative and epistemological category 'fable' in another article by Chakrabarty called 'Humanities in the Anthropocene: The Crisis of an Enduring Kantian Fable' (2016). As we know, Chakrabarty maintains that the core task of a new historiography is to bridge a divide in historiography originating in Vico's new science. In 'Humanities in the Anthropocene', Kant (not Vico) becomes the emblem of a certain cultural modernity that is both one of the causes of the Anthropocene and an epistemic organisation that must be reconfigured to survive *in* this new, geocultural epoch. 'We need', Chakrabarty maintains, to overcome the Kantian fable by finding 'other perspectives' and inventing new 'stories' that go beyond 'any anthropocentric perspective' (Chakrabarty 2016). This is because the fable in question restricts morality and compassion to humans by dividing the human into two separate domains, the biological body and the cultural soul.

Chakrabarty refrains from accounting for the methodological assumptions underlying the idea that 'the separation of the moral life of humans from their animal life in post-Enlightenment narratives' is 'best studied' by turning to an exemplary tale. In a footnote, however, he cites political philosopher Bonnie Honig. Her manner of relating the narrative genre to the foundation and preservation of *forms of life* provides a key to the methodology:

> [T]he stories fables tell about the founding of a form of life invariably serve as a powerful illustration of the now more *subtle and sedimented* but no less *active processes and practices* that *constitute and maintain* our present, daily.
>
> (in Chakrabarty 2016, 396, n. 38, my emphasis)

Fables thus recount the aetiology of cultural forms of life; they tell stori*es* (in the plural) about the 'founding of *a* form of life' (in the singular) and 'illustrate' present everyday life by revealing the hidden foundation of '*our* present'. Hence, 'fable' depicts both the *origin of a particular cultural order or period* and the main rules, processes, and narrative forces upholding it – such as Vico's division in the myth of modern historiography and how it, in a haunting way, 'constitute[s] and maintain[s]' our historiography.

Chakrabarty and Honig's exposition of the relation between narrative and culture, fable, and form of life, represents a particular manner of construing fable or myth. Drawing upon Andrew Von Hendy's history of myth as a scholarly category, we could align it with what he calls a folklorist construal of myth (that later slipped into social anthropology; Von Hendy 2002). This views myth as both a point of access to a form of life *and* expressive of the rules upholding and serving as the charter for a local form of life, i.e. the 'subtle and sedimented but no less active processes and practices that constitute and maintain our present, daily'. As we shall see, this way of construing fable or myth as the 'key' to a particular, local culture or a period in time, is in play in Vico's own negotiations with natural history.

Chakrabarty does not elaborate upon the exemplary status of the Kantian fable in relation to the epistemological exemplarity associated with Vico. Indeed, in 'The Climate of History: Four Theses', it is Vico's division of history into natural and cultural parts that serves as the origin myth or fable, and the persistent paradigm of 'modern' thought on cultural and natural time. As stated, my aim is to complicate the fable of origin of Vico's division by shifting to another historiographical *scale*, to a textual and cultural micro-history. Hence, I will zoom in and slow down the tale of a cultural origin associated with Vico in the historiography of historiography. In particular, I will examine how Vico's concern with fables can be related to a complex art of generic boundary maintenance – between various forms of *historia*.

Verum-factum and authored nature

We observed that Chakrabarty linked the verum-factum principle with the invention of something similar to 'culture' as the true, epistemic domain of history. Accordingly, the task of a historiography able to cope with the dire challenges of the Anthropocene is to bridge a divide in historiography originating in Vico's new science – that is the fable still illuminating our present historiographical practices. Chakrabarty cites two philosophers of history, R. G. Collingwood and Benedetto Croce, as promulgators of Vico's divide – and in particular, of the idea that historiography should restrict itself to the kind of beings with an *interior* that the historian can access. If intentional, conscious actors cannot be identified as the authors behind meaningful events, such events and actors must be left out of history; '[T]he events of nature are mere events, not the *acts of agents* whose *thought* the scientist endeavors to trace' (2009, 202–203, my emphasis). Hence, there are two kinds of event: what we can call authored events that can be traced back to an intention and unintended events without any semantic value at all.

Vico had begun to spell out the historiographical implications of the verum-factum principle in the *Diritto Universale* (1719). Most famously, however, he used it in the *Scienza nuova* of 1725 and 1730/1744[5] to separate what first appears to be human from natural history:

> But in the night of thick darkness enveloping the earliest antiquity, so remote from ourselves, there shines the eternal and never failing light of a truth beyond all question: that the world of civil society [*mondo civile*] has certainly been made by men, and that its principles are therefore to be found within the modifications of our own human mind. Whoever reflects on this cannot but marvel that the philosophers should have bent all their energies to the study of the world of nature, which, since God made it, He alone knows; and that they should have neglected the study of the world of nations, or civil world, which, since men made it, men could come to know.
>
> (Vico 1744, §331; cf. 1725, §40)

As Chakrabarty implies, it appears that 'the world of nature' is excluded from a historiography now solely concerned with human constructions; objects belonging to the category 'culture' – or, in Vico's own vocabulary, *il mondo civile*. According to the constructivist criteria applied in the *Scienza nuova*, knowledge of nature is beyond the epistemic powers of humans because 'we' do not make it. Inversely, 'we' can enter and study the minds of other men, the thought behind their actions, through our participation in a common human mind ('the modifications of our own human mind'). Thus, a modern distinction between the fields of the natural and the human sciences – and the criteria for separating them – appears to be made here.

This, however, erases God from the etiological tale of the history of 'the moderns' and their historiography. In Vico's scheme, there are two intentional agents, God and man, who construct, know, and cause events. Both the natural and the civil world, then, are ultimately regarded as meaningful phenomena – 'authored' by intentional beings. Consequently, nature is not per se devoid of interiority and intentionality, but human cognition has no natural means for attaining knowledge about these aspects of natural phenomena.

Vico's distinction thus assumes an authored cosmos filled with signs. We are accordingly rather far from a secular anthropocentrism as this, for instance, is expressed in Max Weber's definition of culture as 'a finite segment of the *meaningless* infinity of the world process, a segment on which *human beings* confer *meaning* and significance' (Weber 1969, 81, my emphasis; cf. Ødemark 2011). Here, the opposition between *meaning* and *non-meaning* is a homologue to the culture-nature opposition, and thus prescribes two different scholarly approaches to the world: a hermeneutic quest for meaning in the human sciences and a quest for causes in the natural sciences. In Vico, however, nature is not inherently meaningless; it is also an authored product of an intentional agent, but its *events,* and the author behind them, are sublimely beyond *natural* human cognition (human cognition not assisted by divine revelation). In the *Scienza nuova,* both the natural and the civil world are, in the last instance, also *authored* and signifying phenomena. The fact that the Universal Flood is the universal, chronological yardstick in all the different versions of the *Scienza nuova* demonstrates this.

Fables in natural history

We shall return to this consideration and the religious normativity governing Vico's definition of the subject matter below in the context of his quarrel with Burnet and the issue of the dating of the deluge. First, however, we need to explore the genealogy of Vico's version of the verum-factum. This will take us back to what Vico regarded as the primary concern of universal history, fables, and myths, i.e. to the culturally different construal of fable from which Chakrabarty constructs his own fable.

Vico's first version of the verum-factum principle occurs in *On the Ancient Wisdom of the Italians* (1710).[6] In his autobiography,[7] Vico retrospectively scrutinises his own intentions behind the search for 'ancient wisdom'. In this reflection, Francis Bacon's *The Wisdom of the Ancients* is both a model and a contrast. When writing about the wisdom of the ancient Italians, Vico still considered it possible that the fables and myths of ancient nations were vehicles of wisdom and a *true insight into nature*. Curiously, this is an option even Bacon himself appears to be entertaining in the much-commented introduction to the *Wisdom of the Ancients*. 'I do certainly for my part', Bacon says,

> [I]ncline to this Opinion that beneath no small number of the fables of the ancient poets there lay from the very beginning a mystery and an allegory. It may be that my reverence for the primitive time [*prisci seculi*] carries me too far, but the truth is that in some of these fables, as well as in the very frame and texture of the story as in the propriety of the names by which the persons that figure in it are distinguished, I find a conformity and connexion with the thing signified, so close and so evident, that one cannot help believing such a signification to have been designed and meditated from the first.
>
> (English cited from Garner 1970, 268–269; Latin from Bacon 1829, 272 and 276)

Hence, Bacon observes an intentional construction of 'conformity' between the mythical signifier and the signified. This 'connection', moreover, appears to be a mysterious or allegorical way of evoking nature itself. In fact, Bacon explicitly ponders whether ancient fables and myths are a source of insight into the 'nature of things' (*res ipsas*) or the 'culture' of a particular period of time (*antiquitatem*):

> Upon the whole I conclude with this: the wisdom of the primitive ages [*sapientia prisci seculi*] was either great or lucky [*aut magna aut felix fuit*]; great, if they knew what they were doing and invented the figure to shadow out the meaning; lucky, if without meaning it or intending it they fell upon matter which gives occasion to such worthy contemplations. My own pains, if there be any help in them, I shall think well bestowed either way: I shall be throwing light either upon antiquity or upon nature itself [*Aut enim antiquitatem illustrabimus, aut res ipsas*].
>
> (English cited from Garner 1970, 268–269; Latin from Bacon 1829, 272 and 276).

Vico sharply opposed the view that 'ancient fables' could 'be throwing light upon nature itself', for him *fables are clearly a source to antiquity,* that is, to a *particular* historical period and its culture, not to universal nature. Thus, at best, the 'wisdom of the ancients' was an expression of pre-reflective agents who 'without *meaning* it or *intending* [...] fell upon matter which gives occasion to such worthy contemplations'.

This reorientation away from the possibility that Bacon still entertains, that fables are a source of 'nature itself', further informs Vico's quest for a *natural* language – a *favella naturale* – in the later *Scienza nuova*. Here fables, myths, and hieroglyphs are not only delegated to culture and away from Bacon's nature but are attributed to the *primitive beginnings* of culture and language, i.e. to a particular time and place, and the natural connection between signs and things is of the most elementary kind:

> Mutes make themselves understood by gestures or objects that have natural relation [*naturali rapporti*] with the ideas they wish to signify. This axiom is the principle of the hieroglyphs by which all nations spoke in the time of their first barbarism. It is also the principle of the natural speech which Plato [...] guessed to have been spoken in the world at one time. [...] This natural speech [*favella naturale*] was succeeded by the poetic locution of images, similes, comparisons, and natural properties.
> (1744, §225–227)

The collocation of fable and hieroglyph, which Vico here construed as a first, natural language, was an early modern commonplace.[8] The collocation also played a crucial role in early modern natural history. In the wording of William B. Asworth Jr. – the historian who coined the seminal term 'the emblematic worldview' to sum up this paradigm of natural history – 'the Aesopic corpus became an important source' for natural history: 'No student of the peacock would want to ignore the fable of Juno and the peacock' (Asworth 2003, 138). This was because the fable also reflected *generic* human morals shared with animal species (as the Aesopic corpus, and animal fables as a genre still do): 'the peacock complains that he does not have a voice like the nightingale, because there is a moral here for those who are not content with their station in life' (Asworth 2003, 138). Accordingly, nature should be studied together with (cultural) expressions such as fables and hieroglyphs, which also offer real insight into nature, and form a visual language in which 'animals were living characters in the language of the Creator' (Asworth 2003, 137). Using Bacon's phrasing, we could say that fables and hieroglyphs showed a 'conformity and connexion with the thing signified', not the distorted and regional worldview of particular culture but the universal truth about nature.

The constellation of fables and hieroglyphs as a source to universal wisdom and the nature of nature 'itself' had also been theorised by authors such as the Jesuit Athanasius Kircher even after Bacon's time (Marrone 2002).[9] Kircher, one of the most influential intellectuals in the seventeenth century, even regarded 'the science of divinity and of nature' encoded in fables and hieroglyphs as a prefiguration of the trinity. We note that there is nothing *provincial* about Egyptian 'culture' here; on the contrary, their signs provide access to universal nature:

> The Egyptians in fact were the first peoples to celebrate the deeds of gods and the veil of fables, for which reason I dare to assert that the

> hieroglyphic wisdom of the Egyptians was nothing but *the science of divinity and of nature,* presented under various fables and allegorical fictions *of animals and of other natural things,* so that one can say that nothing comes closer to hieroglyphics than the creation of fables and moral tales.
> (cited in Cantelli 1976, 60, my emphasis)

Vico, then, strongly opposed the view that fables and hieroglyphs could serve as sources to knowledge about nature (cf. 1744, §605 on Kircher). In this sense, it is true that Vico leaves *natural history,* but the natural history left behind is also a particular kind of natural history, namely the one that we (applying Asworth's shorthand) can associate with the emblematic worldview. Vico actually continues working with elements or 'symbolic forms' taken from this 'worldview' – namely, 'the hieroglyphs by which all nations spoke in the time of their first barbarism' (1744, §226) We will see this, and how Vico navigates between various subgenres of natural history, in the polemic with Burnet.

Converging natural and cultural history

As noted, the verum-factum principle maintains that 'we, humans, could have proper knowledge of only civil and political institution because we made them, while nature remains God's work and ultimately inscrutable to man' (cf. Chakrabarty 2009). In this rendition, a third genre of historiography is actually implicit, for in addition to the binary distinction between nature and culture, we have sacred history, the story of God's creation. Vico's epistemic distinction could thus be seen as an articulation of a tripartite system of historiographical genres – *historia naturalis, historia civilis* and *historia sacra* – which he uses in several places when dealing with the deluge. Actually, Vico repeatedly cross-references his evidence from these separate genres of history at the most decisive turning points in his new narrative about the origin of human 'culture'. From the *Diritto universale* on, Vico consistently begins with the first *natural speech* and the first fables – seen as rude and primitive phenomena – and the construction of sub-human *bestioni.*

The rebuttal of Burnet occurs in a section of the *Scienza nuova* of 1725 where Vico describes what he calls 'a new critical art' able to make 'obscure' and 'fabulous history' readable (Vico 1725, §93). Again, then, Vico's concern is the reading of fables – and the fable is construed (circularly) as a system of signs and thinking *characterising* a specific period of time *and* the phenomena that explains the period in question. Like the story about the nature-culture distinction as the root and determining logic of modernity, it serves both as the identifying trait of the period and its explanation.

Vico works out what he calls his 'new art of reading' early history in five steps. The first is the presentation of testimony claimed to be *synchronous* with the time when the gentile nations were born (§94–95). The 'uniformity' of 'fabulous traditions' [*tradizioni favolose*] from various peoples shows that the traditions originated simultaneously with the gentile nations as an expression of a natural law. The 'nefarious', 'lascivious' and 'filthy' character of these 'fabulous traditions', however, also sharply separates what Vico calls 'the origins of sacred [*storia sacra*] and profane history [*storia profana*]' (Vico 1725, §94–95). If a distinction between nature and culture, or between natural and cultural history, is crafted in Vico, it is clearly secondary to the religious distinction between true and false religion.

The second step confirms the impression of a theological dominance: it involves what Vico calls 'certain kinds of medals [*certa spezie di medaglie*]' belonging to the first peoples, with which the Universal Flood is demonstrated (Vico 1725, §96–99). The medals in question are actually the 'fabulous traditions' that Vico assumes are characteristic of the first phases of gentile history, i.e. the 'gestures or objects that have natural relation with the ideas they wish to signify', which, as we saw above, 'is the principle of the hieroglyphs by which all nations spoke in the time of their first barbarism' (cf. citation above). The second step, moreover, is also the immediate context of what Vico referred to as the 'dissolution' of Burnet's history of the Earth. Vico's theories of fables, then, will be mobilised against the *Telluris Theoria Sacra* – and thus also to do polemical work in natural history.

The third step concerns 'the *physical* demonstration of the giant'. This proves that the figure of the giant is the origin of 'profane history', 'and that profane history is in continuity with sacred history', which in the last instance explains the nature of the giants (Vico 1725, §100–103). As Rossi maintains, the distinction between the giants (whom Vico, in line with the tradition, saw as inherently related to the deluge) and the Hebrew was biocultural; the 'entire original race' was 'divided into two species', one of giants and the other of men of normal stature (Rossi 1984, 177).

Again, notions of nature and physics appear to be involved in the establishment of this new set of evidence concerning the giant and the beginning of *gentile* history. '[P]roofs can be provided by demonstrations taken from physics, as in the following proof concerning the nature of the first nations' (Vico 1725, §100). Hence, *historia naturalis* and *historia sacra* are seamlessly woven together here. Moreover, 'nature' also occurs in another sense in this last citation. This is because the 'physical nature' of the giant is also mobilised to say something about the '*nature* of the first nations [delle *natura* delle prime nazioni]' (Vico 1725, §100). Thus, from the physicality of the giant, Vico – apparently without any need for translation or mediation between nature and culture – moves to another object: namely, the nature of the 'cultural' subject matter of the *Scienza nuova* – the nature of the nations and the natural law of the nations.[10]

Daston has claimed that it is here, in the discourse on natural law, that 'nature' was first stripped of its relation to a theocentric semiotic: 'Here, and *here alone*, nature was largely emancipated from God' (Daston 1998, 167, my emphasis). It is therefore curious that Vico, as a cultural hero of the modern human sciences and historiography, does not take part in such secularising but instead struggles to realign the natural law of human society with the cure and care of providence. For Vico, it was the sign of providence that brought wild men back into society after the flood.

Demonstrating the deluge

Let us return to the second step and the 'medals belonging to the first peoples, with which the Universal Flood is demonstrated' (Vico 1725, §96–99). These monuments speak to a specific subject matter; they are evidence of a particular theme that Vico treasured, namely the common customs of the *primi popoli*, the human common senses from which natural law was reborn after the deluge. We need to read the long passage within which Vico debunks Burnet:

> And as public medals are the best ascertained documents of certain history, so, for fabulous and obscure history, a few surviving marble remains must take their place as the public medals of the first peoples and as proof of their common customs, of which the following is the most important. A poverty of words of settled meaning lead all the first people to express themselves by means of objects. At first these must been [sic] [natural] solid objects but later they were carved or painted objects, as Olaus Magnus stated in his account of the Scythians and Diodorus Siculus in the writing he left about the Ethiopians. We certainly have the hieroglyphics of the Egyptians, which are depicted on their pyramids, but other fragments from antiquity, with characters of carved objects of the same sort as magical characters of the Chaldeans must first have been, are everywhere to be found. The Chinese also, who vainly vaunt an origin of enormous antiquity, write in hieroglyphics, which goes to show that they originated no more than four thousand years ago. This is confirmed by the fact that, because they remained closed to foreign nations until a few centuries ago, they have only some three hundred articulate words with which, by articulating them in various ways, to express themselves. [...]
>
> This poverty of articulate words in the first [gentile] nations, which was common throughout the universe, proves anew that the Universal Flood occurred before them. And with this demonstration we provide also a true dissolution of the capricious dissolution of the earth dreamt up by Thomas Burnet [*La quale dimostrazione veramente resolve la capricciosa risoluzione della terra immaginata da Tomasso Burnet*], a fantasy that originated first with van Helmot, from which it then passed into Descartes' *Physics.* According to this account, the Flood dissolved the Earth in the south more than in the north, hence the north retained more air in its bowels and, being more buoyant, remained on a higher plane than the south, which then sank into the ocean, causing the earth to decline somewhat from a parallel plane to that of the sun. [But our thesis enables us to refute this] because [had there not been a poverty of articulate words among the gentile nations after the Flood], Idanthyrsus, the king

> of Schytia, would not have replied in hieroglyphics when Darius the Great sent his men to declare war on him.
>
> (Vico 1725, §96–98)

The 'surviving marble remains' from fabulous and obscure history about the 'poverty of words' is primarily inserted in the *Scienza nuova* in an attempt to answer the question about why the study of myth and fable has been unfruitful – until Vico. This 'marble foundation' is then reiterated in almost identical terms towards the end of the paragraph ('poverty of articulate words') – now to refute 'the capricious dissolution of the earth dreamt up by *Tomasso Burnet*'.

This 'dreamt up' Earth was originally pure, a Platonic original of a 'fallen' copy. It had a smooth surface, and what Burnet called the 'wild, vast, indigested heaps of Stones and Earth' that had broken through the Earth's crust after the deluge served as a geological reminder of the sins of men (quoted in Cohen 1996, 49). The deluge even forced the Earth into its present position in space. The asymmetrical distribution of landmasses, water and air after the flood caused the displacement of Earth's axis (Rossi 1984, 107; Cohen 1996, 35–36). Because of this, the Earth tilted to its present angle of approximately twenty degrees (Gould 1987, 35–36). In this geo-theological framework, Eden was located on a perfect Earth, while the 'heaps of stones and earth' were signs – performed a semiotic function – inside the framework of salvation history, or sacred history, as also indicated by Burnet's title (Gould 1987, 32). Hence, like Vico's giants, natural bodies and physical formations were also semiotic entities, signs of the *historia sacra* that scholars should indeed 'endeavour to trace' as authored messages.

Rossi maintains that 'Vico accurately grasped the rigidly mechanistic nature of Burnet's doctrine and, at the same time, refused the image of a Flood that had occurred in very remote times' (Rossi 1984, 107). Hence, Rossi infers that the issue of the historiographical timing of the flood is the most significant for Vico. There is a local context for Vico's use of his own theory of fables and primitive language against Burnet. Vico's scholarly compatriot in Naples, Domenico d'Aulisio – a lecturer in law and, according to Vico, 'a man of universal knowledge' (Vico 1744) – had already condemned Burnet for being a Cartesian, and for promoting ideas about how 'the world had been formed by mere mechanical causes, without divine creation and intervention (Rappaport 1997, 148; cf. Rossi 1984, 73–75). In d'Aulisio's own wording, the atheistic inference to be made from the *Telluris Theoria* was that 'the laws of motion have mechanistically produced the world out of matter, and what is in the world has thus arisen without architect' (d'Aulisio in Rossi 1984, 74). In terms of the verum-factum, there would be no divine making and truth to oppose to human making and truth, and nature would be un-authored and 'meaningless', as in Weber's definition of a secular notion of 'culture'.

The 'threat' to the temporal or chronological ideas that Vico saw as sustaining sacred history came as much from culture and proto-cultural inquiry of past and present 'others' as from burgeoning 'natural science'. Moreover, salient scholarly protocols or paradigms for discovery of evidence and its description – 'best practice' – were shared between disciplines, which we (with hindsight) could distribute according to the modern division.

Stones and antiquarian protocols

The problem Vico sets out to solve in the new critical art is the lack of evidence for the periods before 'certain history'. Vico takes public medals, the best evidence there is for *certain* history, as his point of departure; there 'must' be some 'traces or vestiges' (*vestigi*) that can *take the place* of these public medals and thus furnish similarly good evidence for the fabulous and obscure ages of history. Hence, we get the following analogy:

1 'Just as' [*E siccome*] public medals are the most certain [*più accertati*] documents of 'certain history',
2 so [*così*] there must be some kind of evidence that can *fill the same place* for – fabulous and obscure history.

The lack of public medals and other kinds of evidence – indeed, the very lack that actually makes certain periods of historical time 'obscure' and 'fabulous' – can, Vico asserts, be supplemented with some 'vestiges remaining in marble'. Evidence for the obscure and fabulous ages should consequently share some of the *material* properties of monuments and public medals – should be heavy and durable like stones.

Vico asserts that the 'most grave' [*gravissima*] of the evidence that substitutes for the 'public medals' of 'certain history' is that 'the first peoples spoke with things'. Hence, natural objects were hybrids of *res* and *verba*, a primordial and natural *favella* [*favelle articulate*] before the objects were emptied of matter and became formal representations of things/bodies (in paintings and engravings). Next, this history of signs is turned against the historiographical conceptions of 'foreign cultures', like 'the hieroglyphics of the Egyptians'. Vico then adds information from ethnographic accounts of the present to his authorities on ancient 'others':

> The Chinese also, who vainly vaunt an origin of enormous antiquity, write in hieroglyphics, which goes to show that they originated no more than four thousand years ago. This is confirmed by the fact that, because they remained closed to foreign nations until a few centuries ago, they have only some three hundred articulate words with which, by articulating them in various ways, to express themselves. This demonstrates both the length of time and the great difficulty that nations had to endure before they could furnish themselves with articulate languages [...]. Meanwhile, in our most recent times, travellers have observed that the Americans write in hieroglyphics. This poverty of articulate words in

> the first [gentile] nations, which was common throughout the universe, proves anew that the Universal Flood occurred before them.
>
> (Vico 1725, §97–98)

The aim is clearly to disclaim the chronology of the Chinese (in the present) and the Egyptians (in the past), to cut them down to the correct (temporal) size by associating them with the primitives of the New World – a size measured with the temporal framework of Biblical history as the yardstick. The geological errors of Burnet can be tackled with the same intellectual means as the 'chronological delusions' of the gentile nations, the accuracy of which had been discussed for a long time in Europe (Rossi 1984). Moreover, the inscription of both past and present others in the same category, as nations founded upon fables, demonstrates that these others have not attained the status of 'autonomous' cultures but are condemned with reference to religious criteria.

If Vico employs the rhetoric of novelty, his quest for new kinds of evidence takes him into a conceptual and scientific territory already cultivated by antiquarianism. Here the 'public medals' already played a privileged epistemic role. In the seminal 'Ancient History and the Antiquarian', the historian of historiography Arnaldo Momigliano relates Vico to the division of history into two distinct genres, history and antiquarianism.

> Very conversant with the linguistic, theological and juridical learning of his age, he was practically untouched by the methods of Spanheim, Mabillon, and Montfaucon. He admired Mabillon, and refers at least once to Montfaucon, but did not assimilate their exact scholarship. He was isolated in his times partly because he was a greater thinker, but partly also because he was a worse scholar than his contemporaries. The antiquarian movement of the eighteenth century passed him by.
>
> (Momigliano 1950, 305–306, my emphasis)

Vico, however, does not entirely disregard the protocols of antiquarianism, for he is at least using them to invest fables and hieroglyphs with a status similar to that of public medals. Naturalists also privileged this type of evidence. As historian of geology Rhoda Rappaport observes, '[t]hat naturalists so often assembled both human artifacts and fossils suggest that […] the two kinds of specimen were mentally linked' (Rappaport 1997, 85). In her study of how antiquarian protocols and mental space were largely shared between historians of the Earth and of men in early modern Europe, Rappaport further observes that so-called monuments and medals were seen as 'especially unbiased sources', and thus turned into privileged categories of evidence in both natural and cultural history (Rappaport 1997, 66, 68–69). Moreover, 'the historian's vocabulary permeated geological texts that refer to fossils as the earth's monuments' (Rappaport 1997, 94). When Vico turns to medals and monuments, then, he is also deploying a paradigm for thinking and constructing evidence in early modern antiquarianism and history – a paradigm that was shared across epistemic genres and even the histories of nature and culture.[11]

Conclusion

This chapter has examined an origin myth associated with Vico in 'The Climate of History: Four Theses'. This examination has demonstrated the persistence traffic between various genres of natural and cultural history, even in the supposed author of the modern distinction between these categories in the philosophy of historiography. Moreover, we have also seen that Vico's view of the gentiles as nations founded upon fables demonstrates that these others have not attained the status of 'autonomous' cultures in Vico. On the contrary, they are condemned with reference to the religious logos – as the later Vico (and Kant) will be with reference to the logos of climate science.

I have re-examined the tale of Vico's division by applying a textual microhistory that assumes that he not only is the founder of a new paradigm, *our cultural origin*, with which we must break to survive in the Anthropocene, but also represents a certain *cultural* otherness. Obviously, the Anthropocene was not an issue for Vico, although he, like Burnet, was fully aware that humans as a *species* were responsible for Earth's destiny – in particular, the planet's partial destruction by the deluge. At the very moment (Western) anthropocentrism is revealed as a lethal cosmological aberration that has made us forget our dependency on nature, humanity actually becomes a cosmological agent capable of both destroying and saving nature and the planet. If, as Chakrabarty asserts, this situation calls for a new relation between natural and cultural history, a detailed account of how 'we got there' in the past can perhaps also point forward to a new mode of convergence between these genres – as well as indicate the shifting historical relations between science and myth.

Notes

1 This article was finished during a stay at the Centre for Advanced Study in Oslo (CAS). Thanks are due to Kyrre Kverndokk, Marit Ruge Bjærke and Anne Eriksen for comments and careful reading.
2 Ideally, this should also be related to Chakrabarty's work on postcolonial historiography. I hope to return to this in a later article.
3 In the Latin Middle Ages and Early Modern thought, 'fable' (from the Latin *fabula* translating the Greek *mythos*) served most of the conceptual functions of 'myth', not least, the capacity to simultaneously reference error and ancient wisdom, 'old wives' tales', and the cherished knowledge of ancient Greeks and Egyptians (Von Hendy 2002, 2–3).
4 Vico has been seen as 'the forerunner, the sage who grasped and expressed many truths of the future' (Mali 1992, 1). He has, for instance, been called 'the true father of the concept of culture' (Berlin 2002), the 'rehabilitator' and discoverer of myth (Mali 1992), as well as a producer of a philosophy of images and signs that escaped Western, logocentric semiotics (Verene 1976; Trabant 2004). On this basis, a range of invention and discoveries of themes, topics and perspectives in the history of the human sciences are attributed to Vico's new science.

5 The so-called first (1725) and second *Scienza nuova* (1730 and 1744). As I quote from both the English translation and an Italian version in cases involving key terminology, citations give the year of original publication (1725, 1744, etc). Moreover, citations of the *Scienza nuova* of 1725 and 1744 indicate paragraphs.
6 *De Antiquissima Italorum Sapientia Ex Linguae Latinae Originibus Eruenda.*
7 *Vita di Giambattista Vico scritta da se medesimo.*
8 It had been naturalised by the practice of publishing Horapollo's *Hieroglyphica* in the same volume as the fables of Aesop (Boas in Horapollo 1950).
9 The hieroglyphic tradition also had a profound impact upon natural history. Cf. 'The effect of the hieroglyphic revival on natural history was immediate and profound. Weasels, cranes, and lions became part of a visual language; they were symbols, but even more, they were Platonic ideas, whose meaning the mind could immediately perceive. Animals were living characters in the language of the Creator, and the naturalist who did not appreciate or understand this had failed to comprehend the pattern of the natural world' (Asworth 2003, 137).
10 Cf. the full title of the *Scienza nuova* of 1725: *Princìpi di una scienza nuova/ Intorno all natura delle nazioni/Per la quale si rituovano / I princìpi di altro Sistema/Del diritto naturale delle genti* (1725, n.p. my emphasis).
11 To distinguish history from fables and forgeries, it was important to identify categories of evidence that could resist the attacks upon historical knowledge that had been made by 'pyrrhonists' such as Bayle, La Mothe, Le Vayer, and Huet. By denying history any 'certainty' whatsoever, their attacks actually threatened to define all history as fabulous (Mora 1998, 53). To clear a safe space from which to defend history, '[c]oins – and monuments of all kinds – [...] remained pre-eminently important for the reassurance they could offer that the past recorded in books really had existed and was not a mere series of fictions wrangled over by partisan historians' (Haskell 1993, 23). Thus, a certain type of material and visual evidence served as relics and repositories regarded as more certain than text.

References

Asworth, William B., Jr. 2003. 'The Revolution in Natural History: Natural History and the Emblematic World View', 130–156. *The Scientific Revolution: The Essential Readings*, edited by Marcus Hellyer. Maiden/Oxford: Blackwell Publishing.

Bacon, Francis. 1829. *The Works of Francis Bacon*, Vol. 11. Ed. Basil Montagu. London: William Pickering.

Bakhtin, Mikhail. M. 1981. 'Forms of Time and of the Chronotope in the Novel'. *The Dialogical Imagination: Four Essays.* Austin: University of Texas Press.

Bauman, Richard, and Charles L Briggs. 2003. *Voices of Modernity: Language Ideologies and the Politics of Inequality.* Cambridge and New York: Cambridge University Press.

Becker, Alton L. 1995. *Beyond Translation: Essays Towards a Modern Philology.* Ann Arbor: University of Michigan Press.

Berlin, Isaiah. 2000. *Three Critics of the Enlightenment: Vico, Herder and Haman.* Princeton, NJ: Princeton University Press.

Burke, Peter. 1997. *Varieties of Cultural History.* Ithaca, NY and New York: Cornell University Press.

———. 2007. 'Cultures of Translation in Early Modern Europe', 7–38. In *Cultural Translation in Early Modern Europe,* edited by P. Burke and R. Po-Chia Hsia. Cambridge: Cambridge University Press

Cantelli, Gianfranco. 1976. 'Myth and Language in Vico'. In *Giambattista Vico's Science of Humanity*, edited by G. D. P. V. Tagliacozzo. Baltimore, MD and London: The John Hopkins Press.

———. 1990. 'Gestualità e mito: i due caratteri distintivi della lingua originaria secondo Vico'. *Bolletino del Centro di studi vichiani*, Anno XX: 78–116.

Chakrabarty, Dipesh. 2009. 'The Climate of History: Four Theses'. *Critical Inquiry* 35 (2): 197–222.

———. 2016. 'Humanities in the Anthropocene: The Crisis of an Enduring Kantian Fable'. *New Literary History* 47 (2–3): 377–397.

Cohen, Norman. 1996. *Noah's Flood: The Genesis Story in Western Thought*. New Haven, CT: Yale University Press.

Daston, Lorraine. 1998. 'The Nature of Nature in Early Modern Europe'. *Configurations* 6 (2). https://muse.jhu.edu/article/8141

Garner, Barbara C. 1970. 'Francis *Bacon*, Natalis Comes and the Mythological Tradition'. *Journal of the Warburg and Courtauld Institutes* 33: 264–291.

Gould, Stephen Jay. 1987. *Time's Arrow, Time's Cycle*. Cambridge, MA: Harvard University Press.

Haskell, Francis. 1993. *History and Its Images: Art and the Interpretation of the Past*. New Haven, CT: Yale University Press.

Horapollo. 1950. *The Hieroglyphics of Horapollo*. New York: Pantheon Books for Bollingen Foundation.

LaCapra, Dominick. 1984. *Rethinking Intellectual History: Texts, Contexts, Language*. Ithaca, NY: Cornell University Press.

Latour, Bruno. 1993. *We Have Never Been Modern*. Cambridge, MA: Harvard University Press.

Lloyd, Geoffrey Ernest Richard. 1990. *Demystifying Mentalities*. Cambridge and New York: Cambridge University Press.

———. 2004. 'Styles of Enquiry and the Question of a Common Ontology'. In *Ancient Worlds, Modern Reflections: Philosophical Perspectives on Greek and Chinese Science and Culture*, 76–92. Oxford and New York: Clarendon Press/Oxford University Press.

Mali, Joseph. 1992. *The Rehabilitation of Myth: Vico's 'New Science'*. Cambridge: Cambridge University Press.

Marrone, Caterina. 2002. *I geroglifici fantastici di Athanasius Kircher*. Pavona, Roma: Stampa alternativa & Graffiti.

Momigliano, Arnaldo. 1950. 'Ancient History and the Antiquarian'. *Journal of the Warburg and Courtauld Institutes* 13: 285–315.

Mora, Gloria. 1998. *Historias de mármol: la arqueología clásica española en el siglo XVII*. Madrid: Editorial CSIC-CSIC Press.

Ødemark, John. 2011. 'Genealogies and Analogies of "Culture" in the History of Cultural Translation – on Boturini's Translation of Tlaloc and Vico in I*dea of a New General History of Northern America*'. *Bulletin of Latin American Research* 30 (s1): 38–55. doi: 10.1111/j.1470-9856.2010.00482.x

Ødemark, John, and Eivind Engebretsen. 2018. 'Expansions'. In *A History of Modern Translation: Knowledge Sources, Concepts, Effects*, edited by Lieven D'hulst and Yves Gambier, 85–90. Amsterdam: John Benjamins Publishing Company.

Puckett, Kent. 2016. *Narrative Theory: A Critical Introduction*. Cambridge: Cambridge University Press.

Rappaport, Rhoda. 1997. *When Geologians Were Historians, 1665–1750*. Ithaca, NY: Cornell University Press.

Rossi, Paolo. 1984. *The Dark Abyss of Time: The History of the Earth and the History of Nations from Hooke to Vico.* Chicago, IL: University of Chicago Press.

Said, Edward W. 1997. 'Conclusion: Vico in His Work and in This'. In Edward W. Said *Beginnings: Intention and Method.* London: Ganta Books.

Skinner, Quentin. 2002. 'Interpretation, Rationality and Truth'. In Quentin Skinner *Visions of Politics*, vol. 1, 27–56. Cambridge: Cambridge University Press.

Stengers, Isabelle. 2011. 'Comparison as a Matter of Concern'. *Common Knowledge* 17 (1): 48–63.

Trabant, Jürgen. 2004. *Vico's New Science of Ancient Signs: A Study of Sematology.* London and New York: Routledge.

Vico, Giambattista. (1719–1721) 1936. *Il diritto universale*, vol. 1–3. In Scrittori d'Italia, editor F. Niccolini. Bari: Laterza & Figli.

———. (1725–1731) 1944. *The Autobiography of Giambattista Vico.* Ithaca, NY and London: Cornell University Press.

———. (1744) 1968. *The New Science of Giambattista Vico.* Ithaca, NY and London: Cornell University Press.

———. (1710–1712) 1988. *On the Most Ancient Wisdom of the Italians: Unearthed from the Origins of the Latin Language: Including the Disputation with the Giornale de'letterati d'Italia.* Ithaca, NY: Cornell University Press.

———. 1990. *Vico opere*, vol. 1–2. Ed. Andrea Battistini. Milano: Mondatori Editori.

———. (1725) 1990. *Principi di una Scienza nuova intorno alla natura delle nazioni per la quale si ritrovano i principi de altro sistema del diritto naturale delle genti.* In *Vico opere vol. 2*, edited by Andrea Battistini. Milano: Arnoldo Mondadori Editori.

———. (1744) 1990. *Princìpi di Scienza nuova d'intorno alla commune natura delle nazioni.* In *Vico opere vol. 1*, edited by Andrea Battistini. Milano: Arnoldo Mondadori Editori.

———. (1719–1721) 2000. *Universal Right.* Amsterdam: Editions Rodopi B.V.

———. (1725) 2002. *The First New Science.* Cambridge: Cambridge University Press.

Von Hendy, Andrew. 2002. *The Modern Construction of Myth.* Bloomington: Indiana University Press.

Weber, Max. 1969. *The Methodology of the Social Sciences.* New York: The Free Press.

Part 4

Conclusion

11 Living the climate change

Marit Ruge Bjærke

Into the black hole

Does it really matter to anyone, apart from the researchers who study time, what temporal understandings of climate change we find in our varied material? Our answer, of course, is 'yes'! Knowing about the different temporalities of climate change does matter. It matters because climate change, while a physical phenomenon impacting the world in the shape of melting ice, rises in temperature, and raging storms, must also be conceptualised through texts, images, and practices. Through the many interactions between climate change and concepts such as crisis, catastrophe, prophecy, or deep time, temporality is always one of the factors structuring the relationship between the phenomenon 'climate change' and its conceptualisations. Therefore, knowledge about this relationship is also one of the keys to managing climate change instead of allowing the concept to swallow everything around us and turn the world into one big, black hole (Chapter 4).

To what practical purposes, then, can the knowledge resulting from this book be put? How might it guide people's approaches to social debates or policy assessments? Here, we present two main answers to this question. First, and definitely most importantly, our findings highlight how paying attention to textual genres and practices other than the scientific ones may help to widen the range of possible approaches to and strategies for dealing with climate change. Understanding how different people relate to and cope with the consequences they expect from climate change is of vital importance to the range of different measures that society as a whole must provide to counter it. Second, our findings have implications for those who wish to communicate scientific information on climate change successfully to those outside the scientific community. The two points overlap, of course. Understanding how different groups of people relate to climate change is also a key to successful communication. In the following, however, we will treat the two separately, trying to make some specific points regarding each.

Discarding climate change dichotomies

So, how does studying different time-spaces, or chronotopes, help widen the range of possible approaches to climate change? As we have stated in Chapter 1, one of our aims in studying climate change genres and temporalities is to show that it is a simplification to represent climate change within certain dichotomies. For instance, several of the chapters in this book discuss the relationship between scientific and vernacular knowledge, demonstrating that rather than being dichotomous, this relationship consists of a number of entanglements. In the aquarium exhibition presented in Chapter 7, for instance, the mitigation of climate change and rising CO_2 levels is turned into a set of everyday actions to make them more available to the visitors. At the same time, however, the environmental problems are presented in such a way that they might easily be ignored if the visitors get too interested in the animals and environments on display. Thus, knowledge about environmental problems is supposedly being made more accessible through the focus on everyday actions, but it is still eclipsed as these parts of the exhibition are disregarded in signs and maps, as well as by visitors who are instead mesmerised by the representatives and scientific representations of nature in the aquarium tank itself. Another instance of entangling the scale of knowledge is found in Chapter 9 on civil servant Christen Pram and his attempts to reconcile local knowledge with the temporalities of his scientific method. 'To be made useful, it [Pram's acquired knowledge] could *not* be separated from its local context, it could *not* be cleansed from social and economic circumstance, and it could *not* be rescaled as universal science', Anne Eriksen writes (page 155). Knowledge is never either local or global, and it cannot be separated from the local contexts and experiences of different groups of people.

Moreover, dichotomies such as the experienced versus the abstract, the immediate versus the long-term perspective, and the local versus the global, are often allowed to produce clusters or chains of equivalence. Concepts such as experience, immediacy, short-term, and local perspectives are grouped together on the one hand, with those of the theoretical, abstract, long-term, and universal perspectives on the other, and the terms assigned to either side act almost as synonyms for one another. Again, however, the analyses in this book show such equations to be false. The climate change weatherlore in Chapter 5 is one case where these equivalence chains are entangled. When people's understanding of local weather becomes connected with global climate change, the weather stories themselves are turned into possible arguments for or against regulating the consumption of fossil fuels. This means that the weather – local, ordinary, and concrete as it is – also becomes global and political. Diane E. Goldstein in Chapter 2 also upends the chains of equivalence. She demonstrates how immediacy may signify both power, as when Donald Trump tweets about the weather, and powerlessness, as when ordinary people's everyday economic struggles eclipse the future danger of climate change.

These chapters show that the vernacular and immediate might just as well be connected with power as with powerlessness, and that the local and concrete might be just as politically important as the global. Thus, we conclude that studying a wide variety of genres and their chronotopes, as we have done in this volume, supplies a tool for nuancing and even dissolving spurious chains of equivalence, and for demonstrating the far more organic connections between the different scales as well. This, in turn, opens up the possibility of new thoughts, strategies, and practices. In Chapter 3, Lone Ree Milkær writes of the concept *gjenkunning* that it 'can be seen as a strategy to make the climate changed future about continuity and not about disruption. It is a way for the activists to gain control and manage the risks of the future' (page 45). This book shows it to be one of many, but that is just our point: there is a variety of such temporal strategies out there, many of them relatively far removed from chronological timescales such as the historical or the geological. They need to be discussed in order to shed light on more of the large variety of solutions necessary to deal with a complex problem such as climate change.

Communicating climate change

During recent decades there has been a lot of research on how to communicate scientific climate change knowledge, as well as on how scientific climate change knowledge *is*, in fact, communicated (Doyle 2011; Crow and Boykoff 2014; Hansen and Cox 2015; Hulme 2009; 2017; Kverndokk and Bjærke 2019). There has also been a growing interest in the addressees of such communication, focussing on the ways different publics interact differently with scientific information on climate change and on how experts typically ascribe certain characteristics to the public they address (Maranta et al. 2003; Irwine et al. 2018). As we stated in Chapter 1, 'climate change' is a concept that moves between science, politics, media, and everyday life, thereby creating meaning of different kinds, and inviting actions and reactions that differ significantly from each other. Therefore, a prerequisite for communicating scientific climate change knowledge more effectively is to understand that there are a number of different ways of making sense of this issue. Rather than simply considering some of these as correct and dismissing others as uninformed or naive, there is a need to appreciate the variety and understand how it emerges.

The chapters in this book investigate material representing a few of the genres where climate change understandings are formed – exhibitions, popular science books, questionnaires, television series, and pamphlets. These different genres represent a variety of sense-making practices, based on different approaches to both space and time, which again lead to different expectations of the future. If environmental organisations, scientists, politicians, and policymakers that communicate scientific climate change knowledge are to predict how their communication is received, they need to be aware of the temporal and spatial scales they are tapping into. Different genres, different understandings of temporality and space, and different conceptions of the future need to be taken into account. Based on the chapters in this book, we will highlight three weak spots that are often present in climate change communication, and which are, at least partly, the result of not considering these factors.

First, there is the danger of moving the long timescales of science uncritically to other genres when communicating climate change. As 'climate change' originated as a scientific concept, describing the consequences of chemical changes in the atmosphere, it is easy to use the same scientific temporal scales when communicating other aspects of climate change as well. Showing the level of CO_2 in the atmosphere to be higher than during the last 50 million years, presenting probable global temperatures in 2100, or indicating that *Homo sapiens* has an inherent capacity to destroy the world, makes for powerful rhetoric and imagery. However, if, for instance, environmental NGOs, concerned scientists, or engaged politicians wish their communication to result in a desire to act, a feeling of empowerment, or even a political decision, it is not necessarily helpful to move the timescales of scientific statements into other genres. Rather, as both Diane E. Goldstein and Marit Ruge Bjærke discuss in Chapters 2 and 8, respectively, temporalities such as geological and evolutionary timescales can lead to a feeling of helplessness, a lack of engagement, or an unfair distribution of responsibility between different social or ethnical groups.

Second, there is the danger of turning everything related to the future into climate change. 'Is climate change one of those "black hole" concepts that suck in everything that come into their orbit?' Camilla Asplund Ingemark asks in Chapter 4 (page 66). Tempting as it may be from a communication perspective to connect climate change closely with dramatic concepts such as crisis, catastrophe, or apocalypse, these concepts already contain certain specific, and often deterministic, notions of the future, which become entangled in the presented narrative. Chapters 5 and 8 also highlight ways in which 'climate change' gradually reaches into new areas. Although through very different processes, both weather and species extinctions have been turned into indications of climate change. Geographer Mike Hulme (2011) has used the concept 'climate reductionism' to describe a form of analysis through which the interactions between climate, environment, and society are reduced to one determining factor, namely climate. Hulme attributes this to an epistemological slippage from climate modelling to other fields that are not so easily modelled. However, we contend that the reduction of the future to climate is as much a result of letting temporal and spatial scales slide between genres. In a communication perspective, this should at least result from conscious deliberation rather than being done unconsciously, and with the result that the future only holds a 'black hole' from which there is no escape.

Third, there is the problem of what we might call 'the temporalities of too late'. Some temporal structures enforce the idea that it is too late to do anything about climate change, again creating a sense of helplessness or detachment. Thus, the temporal structure of a climate change narrative can either promote action or promote sitting down and doing nothing. In Chapter 6, Isak Winkel Holm shows the existence of such a distinction between a prophetic mode and an apocalyptic one. As he writes on page 104:

> The two tones should be approached as two cultural conditions of possibility for the way we are sensing time in a world threatened by disaster which is *not now*. In the prophetic tone, we perceive time as an intrahistorical interval of urgency where ethical human action is possible and called for. If we frame the future disasters as apocalyptic, on the other hand, the moment of action disappears in the flat circle of mythical time.

When communicating climate change, it is definitely good to know in which of these tones one is framing the future.

Fear might be an important incentive for action, as activist Greta Thunberg has claimed (2019). However, a feeling of helplessness is not. Climate change communication must balance on the edge between the two feelings. Careful consideration of which temporal and spatial scales to use is necessary to achieve such a balance. Still, the primary goal of our research is not to understand laypeople better so as to tailor-make information campaigns. The value lies *in* these understandings themselves and in the possibility of deriving new futures from them. To be able to imagine the futures we want, we need these different understandings of temporalities as well as those of the relationship between what is and what could be. We need to be reminded that time is not a stable movement from the past via the present to the future. That whatever black hole climate change might look like, it must be met by different people with different approaches and different understandings.

Let the moment of action begin – and stretch out as far as needed!

References

Crow, Deserai A., and Maxwell T. Boykoff, eds. 2014. *Culture, Politics and Climate Change: How Information Shapes our Common Future*. London and New York: Routledge.

Doyle, Julie. 2011. *Mediating Climate Change*. Farnham: Ashgate.

Hansen, Anders, and Robert Cox, eds. 2015. *The Routledge Handbook of Environment and Communication*. London: Routledge.

Hulme, Mike. 2009. *Why We Disagree about Climate Change: Understanding Controversy, Inaction and Opportunity*. Cambridge: Cambridge University Press.

———. 2011. 'Reducing the Future to Climate. A Story of Climate Determinism and Reductionism'. *Osiris* 26 (1): 245–266.

———. 2017. *Weathered: Cultures of Climate*. London: Sage Publications.

Irwin, Alan, Massimiano Bucchi, Ulrike Felt, Melanie Smallman, and Steven Yearley. 2018. *Re-framing Environmental Communication: Engagement, Understanding and Action*. Background paper, The Swedish Foundation for Strategic Environmental Research (MISTRA). Stockholm: MISTRA.

Kverndokk, Kyrre, and Marit Ruge Bjærke. 2019. 'Introduction: Exemplifying Climate Change'. *Culture Unbound: Journal of Current Cultural Research* 11 (3–4): 298–305.

Maranta, Alessandro, Michael Guggenheim, Priska Gisler, and Christian Pohl. 2003. 'The Reality of Experts and the Imagined Lay Person'. *Acta Sociologica* 46 (2): 150–165.

Thunberg, Greta. 2019. *No One Is Too Small to Make a Difference*. London: Penguin Books.

Index

Note: *Italic* page numbers refer to figures and page numbers followed by "n" denote endnotes.